David McHugh

ESSENTIALS

GCSE Design & Technology
Electronic Products

Revision Guide

Contents

Contents

Plastics

Plastics

Plastics are the most widely used materials in commercial production and are divided into…

- **thermoplastics** – can be heated and shaped many times
- **thermosetting plastics** – can only be heated and shaped once.

Main Properties of Plastics

Plastics have many properties:
- Electrical insulator
- Thermal insulator
- Weather resistant
- Chemical resistant
- Corrosion resistant
- Heat resistant
- Many colours available
- Easily shaped and formed

- Easily injected, blown and formed
- Easily fabricated
- Self-lubricating
- Impact resistance.

Types of Thermoplastics

Acrylic (trade name Perspex)…
- is hardwearing and will not shatter, but easily cracks and scratches
- can be machined, drilled and threaded
- is used for signs, aircraft windows, washbasins and covers for car lights.

High Impact Polystyrene (HIPS)…
- is a light, flexible and strong plastic
- can be cut, drilled, injection moulded, vacuum formed and blow moulded
- is used for toys, boxes, containers and cases for electrical goods.

Acrylonitrile Butadiene Styrene (ABS)…
- has a high resistance to impact
- can be cut, drilled, injection moulded, vacuum formed and blow moulded
- is used for car components, tool handles, cases for telephones and kitchen appliances.

Polyvinyl Chloride (PVC) is a lightweight, hardwearing plastic used for pipes, guttering, bottles and window frames.

Nylon is tough with good resistance to wear. It's self-lubricating and can be machined and used for bearings and gear wheels.

Types of Thermosetting Plastics

Glass Reinforced Plastic (GRP) is polyester resin and hardener mixed together and reinforced with glass fibre. It's used for canoes, boats and car bodies.

Epoxy Resin is resin and hardener mixed together and used for castings, printed circuit boards and surface coatings, and the bonding of other materials.

Timbers

Timbers are divided into…

- **softwoods**, e.g. pine
- **hardwoods**, e.g. oak, mahogany, beech
- **manufactured boards**, e.g. plywood, MDF (Medium Density Fibreboard).

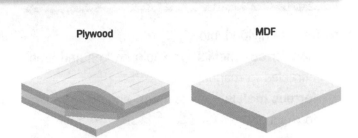

Plywood   MDF

Main Properties of Timber

Grain pattern	Growth rings are visible on the surface.
Colour	Different tree species vary greatly in colour.
Texture	Different tree species have different surfaces and cell textures.
Workability	Some varieties are much easier to work than others.
Structural strength	Different species vary, from weak to very strong.

Types of Timber

Pine is the most widely used timber for general purpose work. It can be cut, machined, planed or nailed without splitting the wood, and joined with screws and PVA glue.

Plywood is constructed from odd numbers (3 or more) of thin layers of wood glued together. The greater the number of layers, the stronger the plywood. Layers of wood are arranged so that their grain directions are turned through 90°. Wood is much stronger along the grain than across it, which gives plywood a uniform strength.

MDF has become very popular for domestic and furniture uses. It's made by bonding wood fibres with resin-based adhesive and is available in a variety of thicknesses. It's extremely useful in school workshops for making moulds for vacuum forming, due to its closely structured inner core.

Quick Test

1. Name the two types of plastic.
2. Name three different plastics.
3. Perspex is the trade name of which plastic?
4. True or false – pine is a softwood.
5. What does MDF stand for?
6. True or false – plywood is constructed from an even number of sheets.

KEY WORDS

Make sure you understand these words before moving on!
- Thermoplastic
- Thermosetting plastic
- Softwood
- Hardwood
- Manufactured board

Metals

Metals

Metals are divided into…

- **non-ferrous metals** – contain no iron and aren't attracted to magnets
- **ferrous metals** – contain iron and are attracted to magnets.

Main Properties of Metals

Elasticity	The ability to regain its original shape after it's been deformed.
Ductility	The ability to be stretched without breaking.
Malleability	The ability to be easily pressed and hammered into shape.
Hardness	Resistance to scratching, cutting and wear.
Brittleness	How likely it is to break without bending.
Tensile strength	The ability to retain strength when stretched.
Toughness	Resistance to breaking, bending or deforming.

Non-Ferrous Metals

Aluminium is soft and ductile, a good conductor of heat and electricity, with a good strength to weight ratio. It's used for cooking foil, window frames, toys and kitchen equipment.

Duralumin is an **alloy** of aluminium with 4% copper and 1% manganese, giving greater strength than aluminium. It's used extensively in the aircraft industry.

Copper is malleable and ductile, a good conductor of heat and electricity with excellent resistance to corrosion. It's used for plumbing and electrical components.

Brass is an alloy of copper with 35% zinc. Brass has good resistance to corrosion, is fairly hard, and a good conductor of heat and electricity. It's used for plumbing, boat fittings and ornaments.

Ferrous Metals

Mild steel is a tough, ductile, malleable metal. It has good tensile strength, poor resistance to corrosion, and is used for general purpose engineering parts.

Mild steel can be joined by soldering, brazing, welding and mechanical fixings. It can be machined by turning, milling, drilling and grinding. It can also be forged, bent and press formed. The outer surface can be case hardened, giving a soft core and a hardened outer layer.

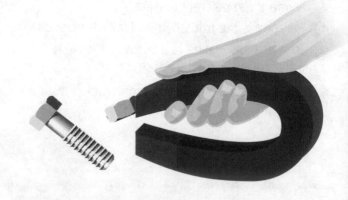

Smart Material	Definition and Uses
Piezo Transducer	• Piezo electric material produces a **small voltage** when it's deformed by **mechanical movement**. It also deforms when voltage is applied. • A commonly used inexpensive transducer is made up of a thin slice of piezo electric material bonded to a brass disc.
Smart Cable	• Uses the **piezo electric principle** for converting **mechanical energy** into **electrical energy**.
Solar Cell (photovoltaic cell)	• Device that converts solar energy into electricity. • Has many applications (e.g. powering calculators, orbiting satellites). • Can be linked together to make solar modules, which in turn can be linked to make larger solar arrays.
Optical Fibre	• Glass or plastic fibre that carries light signals along its length and is used widely in communications, medical applications and hazardous areas. • Used in place of conductive electrical wire because the signal's better with no interference and it isn't as heavy as metal wire.
Thermocolour Sheet	• Made by printing the sheet with a special type of **thermochromic liquid crystal** that changes colour when heated. If touched by hand, body heat causes the material to change colour. • Can also be heated by attaching **resistance wire** to the back of the sheet and used as a heating element when a small electrical current is passed through it.
Liquid Crystal Display (LCD)	• Electronic device used to show numbers or text. The display is made up of **shaped crystals**. • Bar-shape crystals are used in displays that show only numbers, and a pattern of dots are used in displays that show letters and numbers. • Each crystal has an **individual electrical connection**. When an electrical current is passed through the crystal, it changes shape and absorbs light, so the crystal seems darker and visible to the human eye. • LCDs **reflect** and **absorb light** and can't be seen in the dark. A backlight is often fitted to enable them to be seen in the dark, for example, mobile phones.
Smart Wire	• **Shape Memory Alloy** (SMA) that shortens in length when a small electrical current is passed through it. • The most common alloy used to make smart wire is nickel and titanium (often called nitinol).
Quantum Tunnelling Composite (QTC)	• **Flexible polymer** with an unusual electrical property in that it changes from insulator to conductor when compressed. • QTC is expected to revolutionise everyday items like switches and keyboards.

Injection and Blow Moulding

Injection Moulding

Typical materials used in **injection moulding** are **polythene**, **polystyrene**, **polypropylene** and **nylon**:

1. Plastic powder or granules are fed from the hopper into a hollow steel barrel.
2. The heaters melt the plastic as the screw moves it along towards the mould.
3. Once enough melted plastic has accumulated, the hydraulic system forces the plastic into the mould.
4. Pressure is maintained on the mould until it's cool enough to be opened.

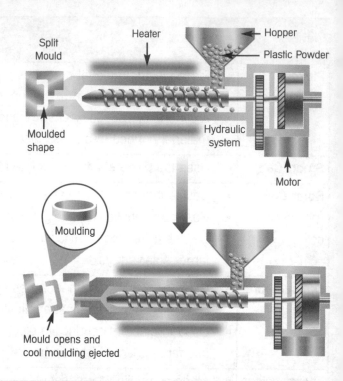

Split Mould

Heater

Hopper

Plastic Powder

Moulded shape

Hydraulic system

Motor

Moulding

Mould opens and cool moulding ejected

Blow Moulding

In **blow moulding** the plastic is forced into the mould to form a tube, which is expanded onto the side of the closed mould with air pressure.

Common materials used in **blow moulding** are **PVC**, **polythene** and **polypropylene**.

Another method of blow moulding uses an injection-moulded bottle blank called a **parison**. This is clamped around the screw thread, heated and blown out to fill the mould. This method is commonly used for drinks bottles as it keeps the bottle neck thicker and stronger.

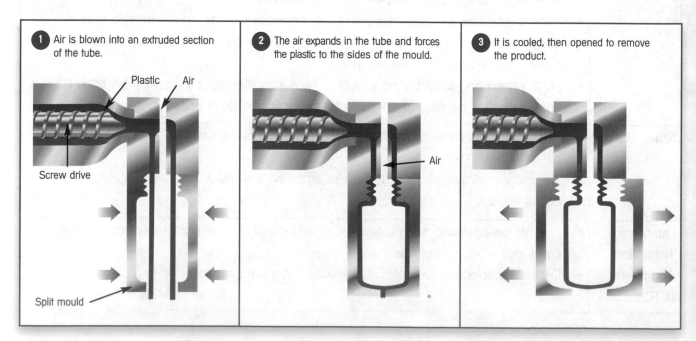

1. Air is blown into an extruded section of the tube.

Plastic

Air

Screw drive

Split mould

2. The air expands in the tube and forces the plastic to the sides of the mould.

Air

3. It is cooled, then opened to remove the product.

Vacuum Forming and Line Bending

Vacuum Forming

Vacuum forming uses **thermoplastic materials** in the form of sheets that can measure up to 1.5m x 1.8m. The most popular material is **High Impact Polystyrene (HIPS)**, which is cheap and easy to form.

The process relies on 'sucking' heated plastic onto the shape of the mould that is required:

1 The plastic is heated and the mould moves close to it. Air is 'sucked out' to form a vacuum.

Heat

Air   Mould   Air

2 Removing the air causes the hot plastic to be sucked onto the mould. As the temperature of the plastic falls, a rigid impression of the mould is formed.

Vacuum

3 The vacuum pump is turned off, allowing air to enter. The mould is lowered, separating it from the final product.

Release

Final product

Air   Mould   Air

Line Bending

Thermoplastic sheets are used in **line bending**, but this time they're heated only along the line of the intended fold by a special heating element.

Temperature switches control the amount of heat produced to cater for different thicknesses of material.

Acrylic sheets are often used for this process, and **bending jigs** can be used to produce accurate angles and shapes.

Remember to keep your fingers away from the heat element and always switch off after use.

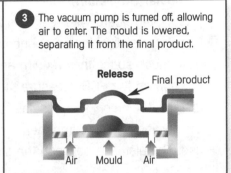

Strip heater

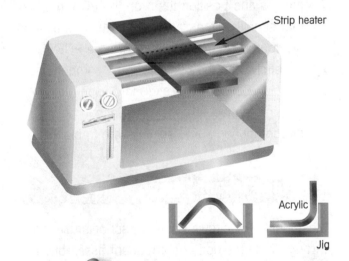

Acrylic

Jig

Quick Test

1. Name two materials used in injection moulding.
2. Name three parts of an injection moulding machine.
3. Name two materials used for blow moulding.
4. What does HIPS stand for?

KEY WORDS
Make sure you understand these words before moving on!
- Non-ferrous metal
- Ferrous metal
- Alloy
- Piezo transducer
- Piezo electric principle
- Polythene
- Polystyrene
- Polypropylene
- Nylon
- PVC
- Thermoplastic material
- High Impact Polystyrene (HIPS)
- Acrylic sheet

PCB Design

Designing A PCB

1 Model or test your chosen electronic circuit to check it works before you start designing the **PCB** (Printed Circuit Board) layout. You could use **Crocodile Technology** or **Circuit Wizard** for **computer simulation** of the circuit, or use a prototyping board with real components.

2 Check the size of the components to determine how much space they require and the best location on the PCB. Measure the size of, and distance between, the pins, leads and connections of the components you intend to use. The distance between the pin connections for ICs is 0.1 inch.

3 Study the **outputs** and **inputs** of each IC. If using more than one IC, look at the relationship between the ICs and position them on the PCB in a way that simplifies the track layout. Remember, there are four ways of placing an IC on a PCB.

4 Carefully plan the position of the components for the PCB design. Make the design big enough to give you enough space to position the components and solder them in place.

5 Draw the components to their actual size on a computer with a 0.1 inch grid. Use vertical and horizontal tracks so that the components are at right angles to each other.

6 Make the **pads** big enough – pads too small with a large hole in the centre make it hard to achieve a good soldered joint. Make your tracks wide enough – narrow tracks can easily be damaged by the ferric chloride solution in the etching tank, or by other processes.

7 Remember that the PCB has a **component side** and a **track side**. You may have to flip your design 180° before you print it onto acetate sheet. Check with your teacher.

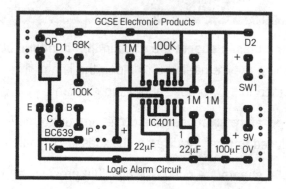

Surface Mount Components

Most electronic circuits made in schools use the **through hole** method of component assembly. In industry, the through hole method has been replaced by **component surface mounting** for the **mass production** of electronic circuits.

In the **surface mount system**, most of the common components have been redesigned without wire leads, meaning that small holes don't have to be drilled in the PCB. **Resistors** and **capacitors** look like tiny bricks with metal end caps, which are placed on the track side of the PCB and soldered in place.

Surface mounting results in smaller and cheaper PCBs and makes maximum use of automatic machinery for the picking and placing of components.

Chip Resistor

Chip Resistor Network

4 Resistors in One Package

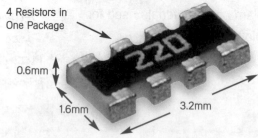

Photo Etch Method

The following process shows you how to produce your PCB using the photographic method:

1. Copy the artwork for the mask onto **acetate sheet** using a **laser printer**.
2. Place the mask in an **ultraviolet light box**. Place a piece of **photo resist board**, cut to the required size, on top of the mask.
3. Leave the photo resist board in the light box for two minutes.
4. Develop the photo resist board for about 30 seconds, using a suitable PCB developer in a ratio of 25 parts cold water to 1 part PCB developer.
5. Wash the photo resist board in cold water.
6. Place the developed board in the **etching tank**, containing **ferric chloride**.
7. Leave the board in the etching tank for about ten minutes.
8. Remove the board from the tank, wash in cold water, then dry.
9. To comply with **health and safety regulations**, always wear goggles, gloves and an apron, and work in a well-ventilated room.

Protoboards

When circuits need developing quickly, **solderless protoboards** (**breadboards**) are used. A protoboard is a plastic case with many holes on top. The leads of components are pushed into the holes and held tightly in place by metal springs, allowing you to build a temporary prototype circuit.

There's a gap down the protoboard's centre for positioning ICs. Protoboards are cheap to buy, use real components and prove that a circuit will work in real life.

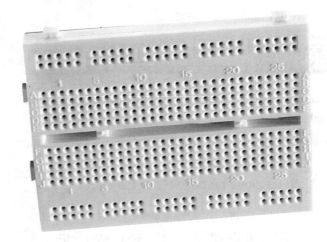

Stripboards

Stripboards (**veroboards**) have components placed on the non-copper side. Component leads are pushed through holes and soldered to the **copper strips** to become a **permanent circuit**. The copper strips run the length of the board. It's sometimes necessary to break the strips with a stripboard cutter to prevent short circuits.

The advantage of stripboard is that it provides a **simple form of PCB**. The disadvantage is that it's quite difficult to use, especially if the circuit is complicated.

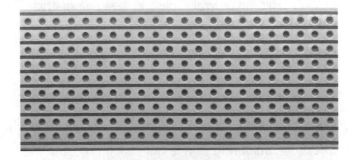

Copper strips run in this direction

PCB Assembly

Quality Control

Inspect your PCB by looking for…

- **breaks** in the tracks
- **missing pads**
- bridging across the tracks and pads.

Look at the **IC socket pads**, using a magnifying glass. If a track might have a hairline break, check the continuity of the track using a multimeter set on the **ohms range**.

Drill the pads using a 1mm diameter drill. Be careful when drilling the pads for ICs, which are only 0.1 inches apart. Safety tip: wear goggles, tuck ties into shirts and tie long hair back.

Quality Assurance

To secure the battery and other devices attached to **flying leads**, drill two 2mm diameter **strain holes** around each relevant pad on the PCB and thread the leads through the holes. The strain holes help to stop the flying leads being pulled out of the soldered joint. This is an example of building **quality assurance** into your product.

Populating the PCB

Most PCBs produced in schools are **single-sided** with holes drilled through the PCB for component location. This means that the components are placed on the opposite side to the copper tracks and pushed as close as possible to the PCB.

Using small pliers, carefully bend the components' leads to make them fit in the gap between their respective pads. The leads of polar components can be insulated in **red** or **black sleeving** to indicate **polarity**. This helps when placing the components onto the PCB and when testing or fault-finding.

Start by attaching the **low profile components**, (e.g. resistors and diodes), then add **taller components** as the circuit progresses. Leave **sensitive components** (e.g. transistors) until the end and always use IC sockets.

Check the **soldering iron** is hot enough by applying **solder** to the tip. Clean the tip with a damp sponge.

Place the iron against the **copper pad** and the lead of the component to be soldered. Hold the iron there for 2–3 seconds, then place the solder on the pad and against the lead and allow the solder to flow onto the pad. The soldering iron and solder shouldn't come into contact. The heat to melt the solder should come from the copper pad and the **component lead**.

This technique ensures that the soldering temperatures of the copper pad, component lead and solder are similar and will help eliminate dry joints. Cut off any unwanted wire. Be careful not to burn your fingers and work in a well-ventilated room.

Soldering Iron

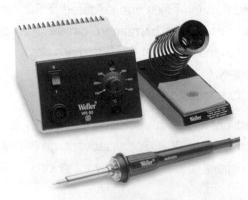

Computer Aided Manufacture

Computer Aided Manufacture

Computer Aided Manufacture (CAM) relies on data known as **machine code**. It is numerical data, which explains why the machinery used for CAM is often called **Computer Numerical Control** (CNC). Drawings are created using **Computer Aided Design** (CAD) packages, so the term CAD/CAM is often used as a single process. Nowadays, machine code is created by the software rather than inputted on a keyboard. This is called **Post Processing**.

Axis Machines

Two-axis machines have two stepper motors controlling movements. One axis controls sideways movement and the other controls front-to-back movement. **Lathes**, **engraving machines**, **plotters** and **vinyl cutters** work this way. These machines are common in schools.

In **three-axis** machines a third axis – up and down – is added which means that more complex machining can take place. **Routers** and **milling machines** fall into this group. These machines are becoming more common in schools and are used mainly for sheet timber and plastics.

The **fourth axis** allows the work to be revolved at the same time as it's being machined and is very much like the addition of a lathe onto a milling machine. This means that full 3D can be achieved in one operation. These machines are very rare in schools but are common in industry.

CO_2 Lasers and Rapid Prototyping

CO_2 **laser machines** cut a wide variety of materials, including fabrics, although school models aren't powerful enough for cutting metals. They can engrave materials such as hard plastics and glass. CO_2 lasers have huge benefits over other CAM systems when cutting as they remove the smallest amount of material with **outstanding accuracy**.

Rapid prototyping is a way of creating full 3D objects direct from a CAD drawing. It builds up layers of wax to make **prototypes**. This is often used in industry and is starting to appear in schools. You can do something similar using layers of paper cut on a vinyl cutter, but it's very time consuming.

Quick Test

1. How many ways can an IC be positioned on a PCB?
2. One side of a PCB is called the component side. What's the other side called?
3. Give an example of quality assurance when making a PCB.
4. True or false – protoboard is a type of PCB.
5. What colour sleeving can be used to indicate the leads of a polar component?

KEY WORDS

Make sure you understand these words before moving on!
- PCB
- Through hole
- Solderless protoboard
- Stripboard
- Computer Aided Manufacture
- Computer Aided Design
- Laser Machine
- Rapid prototyping

Scales of Production

Production Methods

'One off' Production	• One product is made at a particular time. It could be a **prototype** or a very intricate object.
	• It usually takes a **long time**, which results in an **expensive product**.
	• A typical product could be an exhibition display stand.
Batch Production	• A **series of identical products** are made together in small or large quantities.
	• Once made, another series of products may be produced with the same equipment and workforce.
	• A typical product could be a chair.
Mass Production	• The product goes through **various stages** on a **production line** where workers at a particular stage are responsible for a certain part.
	• It usually involves the product being produced for days or even weeks in **large numbers**.
	• This method results in a **relatively inexpensive product**, but production could be halted if a problem occurs at any stage.
	• A typical product could be a car.
Continuous Production	• The product is continually produced over hours, days or even years.
	• This method often results in a **relatively inexpensive product**.
	• A typical product could be screws.
'Just In Time' Production	• Component parts are ordered and arrive at exactly the time the factory needs them.
	• Less storage space is needed, thereby saving on costly warehousing.
	• If the components supply is stopped, the production line stops, which becomes very costly.

Product Life Cycle

Products aren't intended to last forever and have a **life cycle** built into them.

Most design is **evolutionary**. Completely new products are rare and more likely to be adaptations of existing designs. Car designers do this by upgrading existing models. Product life cycle can be seen as **four stages**:

1 **Introduction** – the product is designed, made, tested and launched. Sales are likely to be low and profits don't cover the set-up costs of tooling, wages and materials. The product is made at a loss.

2 **Growth** – sales increase and manufacturing becomes more profitable. Product becomes known by the public and demand rises. Little need to promote the product as it's selling itself.

3 **Maturity** – the product is at its sales peak and its most profitable stage, but it's starting to compete against new and similar products from rivals.

4 **Decline** – in the final stage, sales drop and so does cash flow to the company. The manufacturing costs are higher than the profits so the product is withdrawn or given a facelift to make it sell.

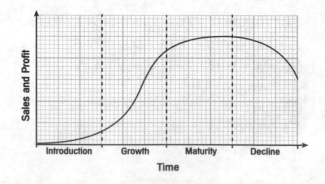

Consumer Protection and Environment

Standards and the Consumers' Association

The **British Standards Institution (BSI)** checks and tests products and components. A standard is an agreed specification for a product or service. The **Kitemark** scheme is an independent and ongoing assessment that the product conforms to the relevant standard. BSI standards are very precise specifications – manufacturers who meet these standards are awarded a Kitemark, which tells consumers that the product has been tested against nationally recognised standards.

The **Conformité Européenne** (CE) symbol tells you that the product meets the minimum requirements from the EU directive to be allowed to be put on sale.

The **Consumers' Association** publishes **Which? Magazine**, whereby similar products are tested and graded against certain criteria, e.g. value for money. Many libraries carry Which? reports.

Kitemark   European Standards

Effects on the Environment

As electronic products become more sophisticated and cheaper to buy, demand increases. Society's increasing reliance on electronic systems that quickly become out of date is leading to a **throwaway society**.

Rapid changes in hardware and software design lead to early obsolescence for mobile phones, televisions, radios and domestic appliances. All raw materials come from this planet, so all waste produced by obsolete electronic products must return to it in some form.

The **recycling symbol** can be found on many products and the packaging used to protect them.

Environmental Checklist ♻

- Will your product have a **detrimental** environmental impact?
- Can you reduce the amount of **raw materials** used and wasted in the manufacturing of the product and its packaging?
- Can you easily **recycle** any materials?
- Can you re-use or recycle materials from other products?
- Can you use **biodegradable materials?**

- Can you reduce the amount of energy used in the production and use of your product?
- Will there be **toxic waste** from your product or the manufacturing processes used (e.g. batteries)?
- Are any parts potentially **hazardous?**

Quick Test

1. Give an example of a product made by batch production.
2. Give an example of a product made by mass production.
3. Give an example of a product made by continuous production.
4. True or false – all products are made to last forever.

KEY WORDS
Make sure you understand these words before moving on!
- Introduction
- Growth
- Maturity
- Decline
- Kitemark symbol
- European Standards symbol
- Recycling symbol

Practice Questions

1 Tick the box of each material to say whether it is plastic, timber & manufactured board, or metal.

	Plastic	Timber & Manufactured Board	Metal
Weather resistant			
It work hardens			
Has elasticity			
Self-lubricating			
Electrical insulator			
Has a grain pattern			
Electrical conductor			
Also called deal			

2 Name three smart materials.

a) ....

b) ....

c) ....

3 Give a safety hazard when using a line bending machine.

....

4 Name two ways that electronic components can be assembled on a Printed Circuit Board (PCB).

a) ....

b) ....

5 Describe what a surface mount resistor looks like.

....

....

6 Give two advantages of using surface mount components.

a) ....

b) ....

7 In industry, how are surface mount components assembled on a PCB?

....

....

8 Give an example of when you would use a protoboard (breadboard).

....

9 List three environmental issues that should be considered when designing a product.

a) ....

b) ....

c) ....

10 Write a brief statement for each stage in a product life cycle.

a) Introduction ....

b) Growth ....

c) Maturity ....

d) Decline ....

11 Draw and label a product life cycle graph on the graph below.

12 List three changes a car designer may carry out when upgrading an existing model in order to make it more attractive to customers and extend the product life cycle. An example has been done for you.

a) Changing the interior of the car.

b) ....

c) ....

d) ....

Electronic Circuit Symbols

Electronic Circuit Symbols

Cell	Battery	Ground	Voltage Rails	Light Dependent Resistor
			+v o— ov o—	

Fuse	Resistor	Variable Resistor	Potentiometer	Thermistor

Push to Break Switch	Push to Make Switch	Single Pole Single Throw Switch	Single Pole Double Throw Switch	Photo Transistor

npn Transistor	Diode	Thyristor	Light Emitting Diode	Flashing Light Emitting Diode

Optoisolator	Tri-colour Light Emitting Diode	Bi-colour Light Emitting Diode	Electrolytic Capacitor	Capacitor

Piezo Crystal Oscillator	Joined Conductors	Crossing of Conductors	Operational Amplifier	AND Gate

OR Gate	NOT Gate	NAND Gate	NOR Gate	EX OR Gate

Lamp	Motor	Voltmeter	Ammeter	Voltage Regulator
				IN OUT REF

Bell	Loudspeaker	Buzzer	Microphone	Field Effect Transistor

Solenoid	555 Timer IC	Single Pole Double Throw Relay	Double Pole Double Throw Relay	Seven Segment Display

Mechanical Switches

Mechanical switches are **hand-operated components** used for making or breaking the flow of electrical current in a circuit:

- **Single Pole Single Throw** (**SPST**) – the simplest type of switch, it has one pole and one contact. It can be thrown into only one contact position and is also available as a **Push to Make Switch** (**PTMS**) and a **Push to Break Switch** (**PTBS**).
- **Single Pole Double Throw** (**SPDT**) – has one pole and two contacts and can be thrown into either of two contact positions.
- **Double Pole Double Throw** (**DPDT**) – has two poles and four contacts and is the same as having two SPDT switches in one package. Can be used for connecting two circuits at the same time or for reversing the **polarity** of electric motors.

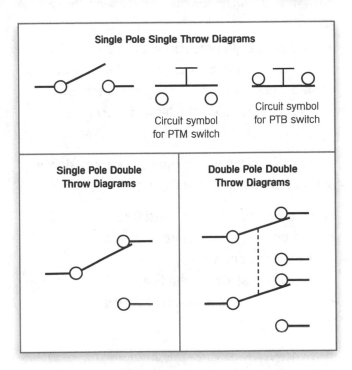

Single Pole Single Throw Diagrams

Circuit symbol for PTM switch

Circuit symbol for PTB switch

Single Pole Double Throw Diagrams

Double Pole Double Throw Diagrams

Types of Mechanical Switches

The table shows some of the different kinds of mechanical switches.

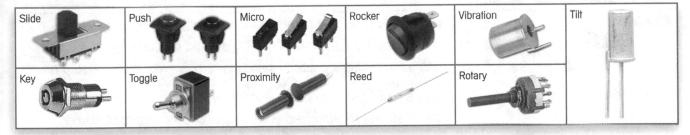

| Slide | Push | Micro | Rocker | Vibration | Tilt |
| Key | Toggle | Proximity | Reed | Rotary | |

Quick Test

1. What does PTMS stand for?
2. What does PTBS stand for?
3. True or false – a battery, LED and buzzer are all polar components.
4. What does SPST stand for?
5. What does SPDT stand for?
6. What does DPDT stand for?

Batteries

Batteries

Batteries...

- are safe, **self-contained** energy sources that convert **chemical energy** into **electrical energy**
- are available in several sizes and voltages, which have different internal properties that make them suitable for particular uses.

Avoid projects that use a **mains electricity power** supply due to the potential **dangers**.

When selecting a battery, consider...

- the **power requirements** of your circuit
- its **voltage** and **type**
- the **physical size** of the battery
- how much you're willing to **spend**.

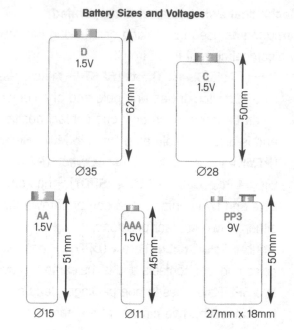

Battery Sizes and Voltages

Battery Types

Zinc Carbon	• The **most basic battery**; cheaper version of alkaline. • Suitable for projects that don't require high outputs; the voltage level **falls** during use.
Alkaline	• **Longer life** than zinc carbon batteries, but more expensive. • Voltage level doesn't fall as quickly as zinc carbon. • Has a much **higher capacity** than zinc carbon; provides more **amp hours**.
Silver Oxide	• Very small 1.5V button cells used in watches, clocks, calculators, cameras and miniaturised electronic products, where size and performance of power source are important. • Provides an **almost constant voltage level** until the cell is discharged.
Lithium (Non-rechargeable)	• Small, expensive, 3V disc cells with a **very long life**. • Used in miniature electronic products where a **consistent**, **accurate power source** is needed. • Used in personal organisers, handheld games consoles, calculators and watches.
Nickel Cadmium (Rechargeable)	• Available in all common sizes. • Retains its voltage level well while charged, but doesn't have the same capacity or working life as alkaline. • Can be recharged many times so it's **cost-effective**. • Used in many appliances, e.g. cordless drills.

Resistors

Resistors are common **non-polar** components in electronic circuits. They're used to limit the amount of current flowing in the circuit and to set the voltage levels in certain parts of the circuit.

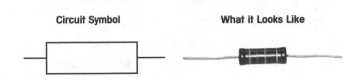

Circuit Symbol   What it Looks Like

The Ohm

The **ohm** (Ω) is the unit of measurement for **resistance**. It is very small so resistors are given the **multipliers** of...
- R to signify times one
- K to signify times a thousand
- M to signify times a million.

The **decimal point** is replaced with one of the multiplier letters R, K or M. For example...
- 1200 ohms is 1.2 kilohms, which is written 1K2
- 1 000 ohms is written 1K
- 1 500 000 ohms is written 1M5
- 820 ohms is written 820R.

The Resistor Colour Code

Resistors are painted with **four bands of colour** that show their **resistance value**. When reading the value, hold the resistor with the single gold or silver tolerance band on the right. This is the **resistor colour code**.

The first two bands give the first two numbers of the resistor's **value**. The third band gives the **number of zeros to be added**. A silver band indicates a tolerance of **±10%** and a gold band indicates a tolerance of **±5%**.

A 1K resistor with a tolerance of ±5%

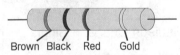

Brown Black Red Gold

A 470R resistor with a tolerance of ±10%

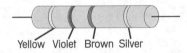

Yellow Violet Brown Silver

A 68K resistor with a tolerance of ±5%

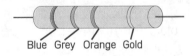

Blue Grey Orange Gold

Colours	Band 1	Band 2	Band 3	Band 4 (Tolerance)
Black	0	0	x1	
Brown	1	1	x10	
Red	2	2	x100	
Orange	3	3	x1 000	
Yellow	4	4	x10 000	
Green	5	5	x100 000	
Blue	6	6	x1 000 000	
Violet	7	7	x10 000 000	
Grey	8	8		
White	9	9		
Silver				±10%
Gold				±5%

Resistors

Preferred Values

Exact values of resistors aren't required in most electronic circuits, so only a certain number of **preferred values** are made to a particular **tolerance** or **accuracy**.

This ensures **maximum coverage** is obtained without unnecessary overlapping, using a limited number of series.

E12 Series	E24 Series
Resistors with a silver tolerance band ±10% belong to the E12 series, which has 12 basic values: 10, 12, 15, 18, 22, 27, 33, 39, 47, 56, 68, 82 Resistors in this series are also available in multiples of 10, 100, 1000, etc. of the 12 basic values. For example, 120R, 1K8, 39K, 5M6, 82M and so on.	Resistors with a gold tolerance band ±5% belong to the E24 series, which has 24 basic values: 10, 11, 12, 13, 15, 16, 18, 20, 22, 24, 27, 30, 33, 36, 39, 43, 47, 51, 56, 62, 68, 75, 82, 91 Again, resistors in this series are available in multiples of 10, 100, 1000, etc. of the 24 basic values. For example, 160R, 22K, 6M8 and so on.

Other Types of Resistor

The resistance value of **potentiometers** (potential dividers) or **variable resistors** is changed by moving the wiper along the **resistance track**.

The **maximum resistance value** is printed on the variable resistor's body.

Component	What it Looks Like
Rotary Potentiometer or Variable Resistor	
Slide Potentiometer or Variable Resistor	
Preset Potentiometer or Variable Resistor	

DIL and SIL Resistors

Resistors are also available in Dual and Single in line networks.

DIL   **SIL**

8 resistors in one package   4 resistors in one package

Current Limiting Resistors

Components such as **Light Emitting Diodes** (LEDs) and **transistors** can be protected by using a **current limiting resistor**.

This limits the current and prevents components from being damaged.

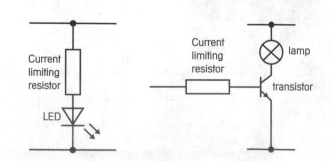

Current limiting resistor

LED

Current limiting resistor

lamp

transistor

Ohm's Law

Ohm's Law is possibly the most important formula in electronics.

Electrical current flows through some materials more easily than others when a voltage is applied. This opposition to current flow is **resistance** and it is measured in **ohms**. Current is measured in **amps** and voltage is measured in **volts**.

Ohm's Law states that the relationship between **current (I)**, **voltage (V)** and **resistance (R)** in an electrical circuit is:

Voltage (V)	=	Current (I)	**X**	Resistance (R)

The formula can be rearranged using the triangle. Cover up the unknown value to read the formula you require:

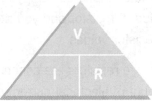

To find V...   To find I...   To find R...

$$V = I \times R \qquad I = \frac{V}{R} \qquad R = \frac{V}{I}$$

Ohm's Law Examples

1 What current passes through a 180R resistor if the voltage across it is 9 volts?

Using the formula triangle:

$$I = \frac{V}{R}$$

$$= \frac{9V}{180R}$$

$$= \frac{9}{180}$$

$$= \textbf{0.05A or 50mA}$$

2 A current of 2mA passes through a 1K resistor. What is the voltage across the resistor?

Using the formula triangle: $V = I \times R$

$V = I \times R$ **OR** $V = I \times R$

$\quad = 2mA \times 1K$ $\quad = 2mA \times 1K$

$\quad = (2 \times 10^{-3}) \times (1 \times 10^{3})$ $\quad = \dfrac{2 \times 1 \times 1\,000}{1\,000}$

$\quad = \textbf{2V}$ $\quad = \textbf{2V}$

Quick Test

1. Name four sizes of batteries.
2. True or false – the PP3 battery is a 12V battery.
3. What is the value of a resistor with the colour bands orange, orange and brown?
4. Why are resistors produced with a certain number of preferred values?
5. Give two differences between resistors from the E12 series and the E24 series.
6. What do SIL and DIL stand for?

KEY WORDS

Make sure you understand these words before moving on!
- Battery
- Resistor
- Ohm
- Resistor colour code
- Potentiometer
- Current limiting resistor
- Ohm's Law

Resistors in Series and Parallel

Resistors in Series

Resistors can be connected together in two ways to give **different total values of resistance** or to **divide voltage or current**. This is useful if you don't have the correct value of resistor and you need to make it from other resistor values.

When two or more resistors are connected in series, the same amount of electrical current passes through each resistor and the total resistance is **equal** to the sum of the separate resistances.

This is the formula for calculating the **total resistance of resistors in series**:

$$R_{total} = R_1 + R_2 + R_3$$

Circuit Diagram of Total Resistance in Series

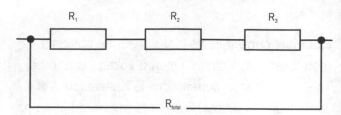

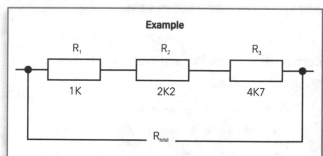

Example

$R_{total} = R_1 + R_2 + R_3$

$R_{total} = 1K + 2K2 + 4K7$

$R_{total} = 7K9$

Resistors in Parallel

When two or more resistors are connected in parallel, the voltage across each resistor is the **same** and the current flowing through each separate resistor is dependent upon the resistance. The total resistance is always **smaller** than the **smallest individual resistance**.

This is the formula for calculating the **total resistance of resistors in parallel**:

$$\frac{1}{R_{total}} = \frac{1}{R_1} + \frac{1}{R_2} \quad \textbf{OR} \quad R_{total} = \frac{R_1 \times R_2}{R_1 + R_2}$$

Circuit Diagram of Total Resistance in Parallel

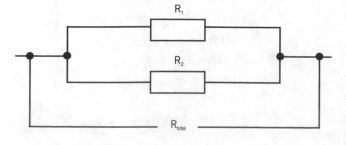

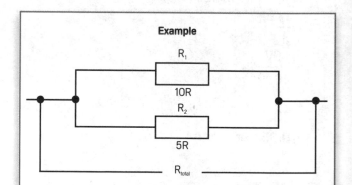

Example

$$\frac{1}{R_{total}} = \frac{1}{R_1} + \frac{1}{R_2} \qquad \textbf{OR} \qquad R_{total} = \frac{R_1 \times R_2}{R_1 + R_2}$$

$$\frac{1}{R_{total}} = \frac{1}{10} + \frac{1}{5} \qquad\qquad\qquad\qquad = \frac{10 \times 5}{10 + 5}$$

$$\frac{1}{R_{total}} = \frac{1}{10} + \frac{2}{10} = \frac{3}{10} \qquad\qquad\qquad = \frac{50}{15}$$

$$R_{total} = \frac{10}{3} \qquad\qquad\qquad\qquad\qquad = 3R3$$

$$R_{total} = 3R3$$

How A Potential Divider Works

A **potential divider** (a **voltage divider**) consists of a **voltage supply** (Vs), which is applied across resistors **R₁** and **R₂**, arranged in series. The voltage across the lower resistor is 'tapped off' to become **voltage out** (Vout). Vout can be calculated using this formula:

$$Vout = \frac{R_2}{R_1 + R_2} \times Vs$$

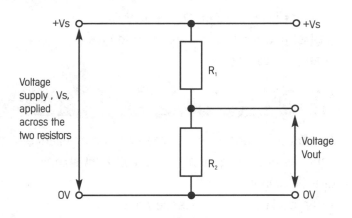

Voltage supply , Vs, applied across the two resistors

Increasing and Decreasing Resistance

If the resistance of either resistor is increased (or decreased), the voltage across that resistor also increases (or decreases).

This alters the **value** of Vout, as seen in the following examples (the ratio between R₁ and R₂ determines the value of the voltage out).

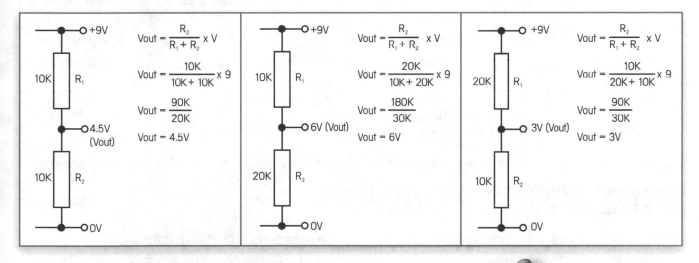

$Vout = \frac{R_2}{R_1 + R_2} \times V$

$Vout = \frac{10K}{10K + 10K} \times 9$

$Vout = \frac{90K}{20K}$

$Vout = 4.5V$

$Vout = \frac{R_2}{R_1 + R_2} \times V$

$Vout = \frac{20K}{10K + 20K} \times 9$

$Vout = \frac{180K}{30K}$

$Vout = 6V$

$Vout = \frac{R_2}{R_1 + R_2} \times V$

$Vout = \frac{10K}{20K + 10K} \times 9$

$Vout = \frac{90K}{30K}$

$Vout = 3V$

Quick Test

1. What is the formula for two resistors in series?
2. True or false – when resistors are connected in series, their total resistance is greater.
3. True or false – when resistors are connected in parallel, their total resistance is greater.
4. True or false – a potential divider is a voltage divider.
5. When Vs = 9V, what size resistors would you choose to give Vout = 4.5V?

KEY WORDS
Make sure you understand these words before moving on!
- Resistors in series
- Resistors in parallel
- Potential divider
- Voltage supply
- Voltage out

LDRs and Thermistors

LDRs and Thermistors

Sensors such as **Light Dependent Resistors** (**LDRs**) and **thermistors** are used to give a changing resistance.

Sensors are often used in conjunction with a **fixed** or **variable resistor** in a potential divider. The variable resistor is used to set the **switching level**.

Light Dependent Resistors

An **LDR** is a special type of resistor. The resistance changes according to the amount of **light** shining directly onto it:

- In **darkness** its resistance **rises** to about 10M.
- In **bright light** its resistance **falls** to around 1K.

The LDR can convert changes in light levels into changes in electric current. LDRs are non-polar.

Circuit Symbol   **What it Looks Like**

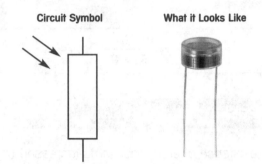

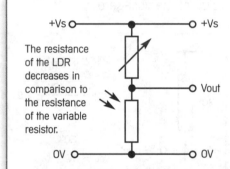

The resistance of the LDR decreases in comparison to the resistance of the variable resistor.

If Vout is high and the light intensity increases...

...then Vout changes from high to low.

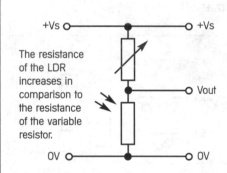

The resistance of the LDR increases in comparison to the resistance of the variable resistor.

If Vout is low and the light intensity decreases...

...then Vout changes from low to high.

Thermistors

A thermistor is a type of resistor. Its resistance changes according to **temperature**.

A thermistor converts changes in temperature into changes in electric current.

The most popular thermistor is the **negative coefficient** type. Its resistance **decreases** as it gets hotter.

Thermistors are non-polar.

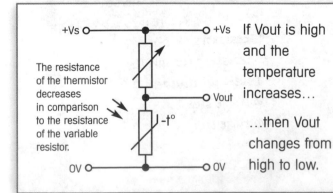

The resistance of the thermistor decreases in comparison to the resistance of the variable resistor.

If Vout is high and the temperature increases...

...then Vout changes from high to low.

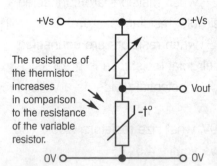

The resistance of the thermistor increases in comparison to the resistance of the variable resistor.

If Vout is low and the temperature decreases...

...then Vout changes from low to high.

Resistor Power Ratings

When a current flows through a resistor, it converts electrical energy into **heat**.

The **rate of conversion** (power, P) can be calculated by **multiplying** the **voltage across the component** (measured in volts, V) by the **current flowing** (measured in amps, I). This gives the **power** (measured in **watts**, W).

$$P = V \times I$$

The formula can be rearranged using the triangle. Cover up the unknown value to read the formula you require:

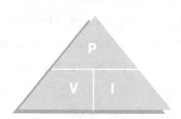

To find V...   To find I...   To find P...

$$V = \frac{P}{I} \qquad I = \frac{P}{V} \qquad P = V \times I$$

Resistor Power Ratings Example

Calculate the power rating of the resistor shown in the diagram. An LED needs 2V across its leads to make it work.

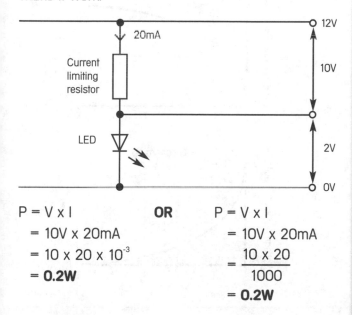

P = V x I	**OR**	P = V x I
= 10V x 20mA		= 10V x 20mA
= 10 x 20 x 10^{-3}		= $\frac{10 \times 20}{1000}$
= **0.2W**		= **0.2W**

Resistors are available with power ratings of 0.125W, 0.25W, 0.5W, 1.0W, 2.0W and 2.5W. The most common size of resistor used in electronics has a power rating of 0.25 watts.

Select a power rating that is at least 1.5 times greater than the calculated size. The larger the power rating, the larger the physical size of the resistor.

Quick Test

1. What is the resistance of an LDR when...
   a) in darkness?
   b) in bright light?
2. What is a negative temperature coefficient thermistor?
3. True or false: P = V x I.
4. What unit is used for power?
5. What is the power rating of the most common resistor used in electronics?

KEY WORDS
Make sure you understand these words before moving on!
- Light dependent resistor
- Thermistor
- Negative coefficient
- Watt
- P = V x I

Capacitors

Capacitors

Capacitors are used in electronic circuits to store an electrical charge. The **capacitance value** and the **working voltage** is printed on the capacitor.

Capacitors…
- create **time delays** (monostable)
- control the **frequency** of **pulse generators** (astable)
- **smooth the input** across a power supply.

The unit of capacitance is the **farad**, which is a large quantity. The farad is divided into smaller units:
- 1 **microfarad** (1 μF) = 10^{-6}F (0.000001 F)
- 1 **nanofarad** (1 nF) = 10^{-9}F (0.000000001 F)
- 1 **picofarad** (1 pF) = 10^{-12}F (0.000000000001 F)

The microfarad and the nanofarad are the two most popular ranges of capacitors.

Charging and Discharging a Capacitor

A capacitor can be **instantly charged** by connecting a power supply across its two leads.

The rate at which a capacitor charges…
- can be controlled by connecting a current limiting resistor in series with the capacitor
- depends upon the sizes of the capacitor and the current limiting resistor – this is known as the **time constant** and is calculated using this formula:

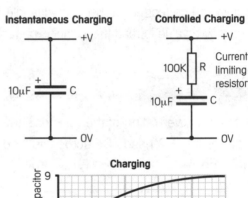

| Time Constant (seconds) | = | C (farads) | X | R (ohms) |

After one time constant, the capacitor is charged to about **0.6** of its full charge and would require a further **four time constants** to **fully charge**.

Capacitors discharge in the same way that they charge and would lose about 0.6 of their charge every time constant.

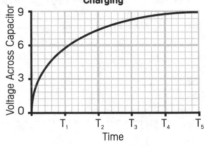

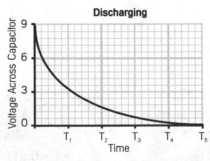

Tolerance of Capacitors

Electrolytic capacitors, are made with large capacitance values to a tolerance of **±20%** and are only available in multiples of 1, 2.2 and 4.7.

Non-electrolytic capacitors, such as **silvered mica** and **polystyrene**, are manufactured to a tolerance of **±1%**. More capacitor values are available.

Electrolytic Capacitor

Non-electrolytic Capacitor

Capacitors in Series and Parallel

Capacitors in Series	Capacitors in Parallel
The total capacitance is always **smaller** than the **smallest individual capacitance**.	The total capacitance is **equal** to the sum of the **individual capacitances**.

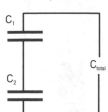

$$\frac{1}{C_{total}} = \frac{1}{C_1} + \frac{1}{C_2}$$

$$\frac{1}{C_{total}} = \frac{1}{100} + \frac{1}{100} = \frac{2}{100}$$

$$C_{total} = \frac{100}{2} = \mathbf{50\mu F}$$

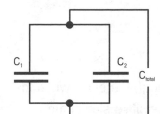

$$C_{total} = C_1 + C_2$$

$$C_{total} = 100\mu F + 100\mu F$$

$$C_{total} = 200\mu F$$

Types of Capacitors

Capacitors can be **electrolytic** or **non-electrolytic**.

Electrolytic capacitors…

- are polar components available with radial or axial leads; the cathode lead has to be connected to 0V and is marked with a stripe on its case
- usually have much **higher values**
- must be connected in the circuit the correct way round.

Non-electrolytic capacitors…

- are normally **smaller** and have a capacitance value of **less than 1μF**
- can be connected either way around in a circuit, so they are **non-polar components**
- normally have **radial leads**.

Polarised Capacitor (Electrolytic)

+

Anode

Cathode

Radial Electrolytic Capacitor

Non-polarised Capacitor (Non-Electrolytic)

Cathode   Anode

Axial Electrolytic Capacitor

Quick Test

1. What do capacitors store in electronic circuits?
2. Give three uses of capacitors in electronic circuits.
3. Why is the farad divided into sub-units?
4. True or false – a non-electrolytic capacitor can be connected in a circuit either way round.
5. What's the difference between a radial and axial electrolytic capacitor?

KEY WORDS
Make sure you understand these words before moving on!
- Capacitor
- Farad
- Time constant
- Electrolytic capacitor
- Non-electrolytic capacitor

Diodes

Diodes

A **diode** is a component that allows current to flow through it in only **one direction**. It's like a one-way street in a circuit.

A diode has **two leads**:
- The **anode**.
- The **cathode**.

A current will only flow through the diode when the anode is connected to the **positive** side of the power supply and the cathode is connected to the **negative** side. When connected this way, the diode is **forward biased**.

Diodes are polar components.

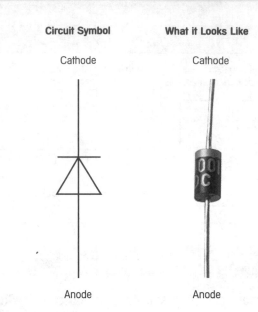

Circuit Symbol | What it Looks Like

Cathode | Cathode

Anode | Anode

How Diodes Work

Diodes are used to...
- protect transistors from back **electromotive force** (emf)
- **protect components** against the possibility of **incorrect battery polarity**
- **direct electronic signals** to **stop feedback** from outputs (steering diodes).

A way to remember if a diode is forward biased is that the arrow showing the direction of the current points in the **same direction** as the diode. This also applies to LEDs.

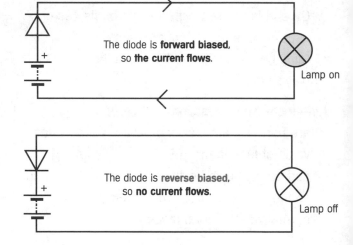

The diode is **forward biased**, so **the current flows**.

Lamp on

The diode is **reverse biased**, so **no current flows**.

Lamp off

Photodiode

A **photodiode** is a special type of diode with anode and cathode leads. Photodiodes are either in a **transparent case**, like an LED, or a **metal case** with a glass window on top.

A photodiode is normally connected in **reverse bias**. When light falls on the glass window, **reverse current** starts to flow and **increases proportionally** to the increase in light.

Metal-cased Photodiode

Light Emitting Diodes

Light Emitting Diodes (LEDs) are a special type of diode. They give out **light** when current passes through them. Like ordinary diodes, LEDs only allow current to flow in one direction.

An LED has two leads – an anode and a cathode. On most LEDs the cathode lead is **shorter** and alongside it, on the plastic case, is a small flat.

LEDs come in several sizes and colours, including **multicoloured**, **flashing** and **tri-coloured** (which have **two** anodes and **one** cathode).

You can limit the current flowing through an LED to **20mA** with a **current limiting resistor**. When calculating resistance, an LED needs about **2V** across its leads to make it work.

When using **flashing LEDs**, there's no need to use a current limiting resistor. LEDs are polar components.

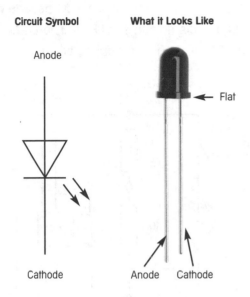

Circuit Symbol   What it Looks Like

Anode

Flat

Cathode   Anode Cathode

Example

Work out the resistance of the current limiting resistor.

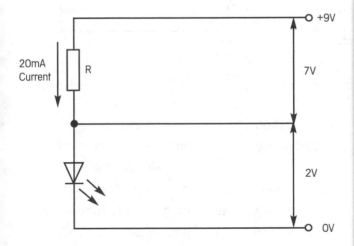

$$\text{Resistance} = \frac{\text{Supply voltage} - 2 \text{ volts}}{\text{Current through LED}}$$

$$= \frac{9V - 2V}{20mA}$$

$$= \frac{7}{20 \times 10^{-3}} \qquad \textbf{OR} \qquad = \frac{7 \times 1\,000}{20}$$

$$= \frac{7}{0.02} \qquad\qquad\qquad\quad = \frac{7\,000}{20}$$

$$= \textbf{350R} \qquad\qquad\qquad\quad = \textbf{350R}$$

Quick Test

1. True or false – a diode is a polar component and only allows current to flow in one direction.
2. What are the two leads of a diode called?
3. Give three uses of diodes in electronic circuits.
4. What does LED stand for?
5. What two features of an LED identify the cathode lead?

KEY WORDS

Make sure you understand these words before moving on!
- Diode
- Anode
- Cathode
- Forward biased
- Reverse biased
- Light Emitting Diode

Diodes

Steering Diodes

Diodes can be used to **direct electronic signals** to stop outputs from **feeding back** to other outputs. The circuit below shows a **4017 IC decade counter** with its outputs driving six LEDs. The LEDs connect to the 4017 IC to give a rippling effect.

The LEDs light in the sequence 1, 2, 3, 4, 5, 6, 5, 4, 3, 2 and so on. LEDs 2, 3, 4 and 5 receive electronic signals from two different outputs, so the circuit requires eight **steering diodes** to stop the signals feeding back into the 4017 IC.

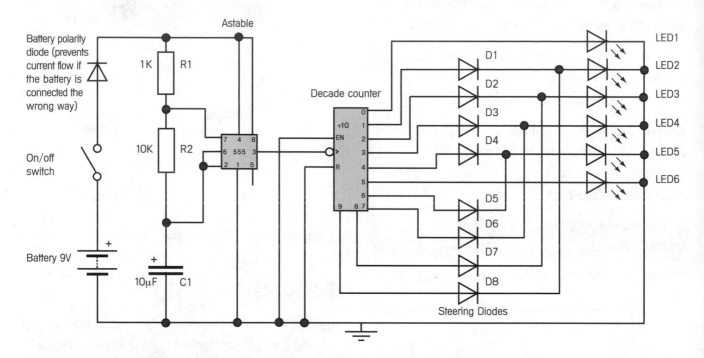

Seven-Segment Display

Most types of counters use **seven-segment LED displays** to show numbers.

Each display contains seven separate LEDs, arranged in a pattern to form a number from **0 to 9** and an **extra** LED for the decimal point. Each segment has a specific letter (**a to g**). By forward biasing two or more segments, numbers 0 to 9 can be displayed.

LED displays come as either a **single digit** or **multi-digit** display, which can contain up to 10 digits. All seven-segment LED displays are either...

- **common cathode**, or
- **common anode**.

A common cathode LED display (the most popular type), has all the cathodes connected together.

The common cathode is connected to 0V while the **individual segment anodes** are each connected to separate current limiting resistors. If a **single current limiting resistor** is used in the common cathode to 0V, a **variation in LED brightness** will occur depending upon the number displayed.

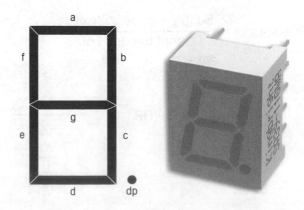

Seven-Segment Latch and Driver 4511 IC

An LED display will use up to **eight outputs** if it's connected to a PIC microcontroller.

By using a 4511 IC, which controls the LED display by **binary code**, only four PIC microcontroller outputs are needed.

Pins 0, 1, 2 and 3 are microcontroller outputs.

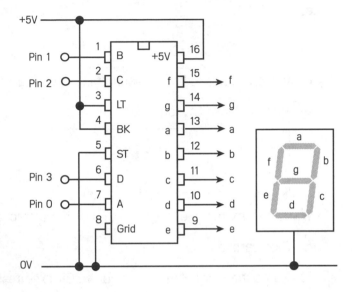

Decade Counter with Decoded Output

A seven-segment display can also be controlled using a 4026 IC, which is a **decade counter with seven-segment decoded outputs** suitable for directly driving LED displays.

This controls the seven-segment display by **decimal code** and only uses two microprocessor output pins.

Pins 0 and 1 are microcontroller outputs.

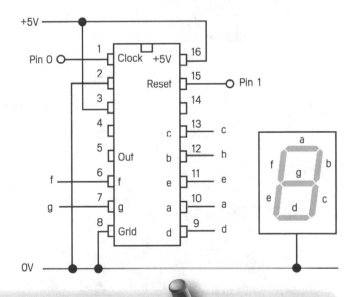

Quick Test

1. What is the purpose of a steering diode?
2. What is the purpose of a battery polarity diode?
3. True or false – when a diode is forward biased, the current doesn't flow.
4. True or false – seven-segment displays contain eight separate LEDs.
5. Why does each LED in a seven-segment display require its own current limiting resistor?

KEY WORDS

Make sure you understand these words before moving on!
- Steering diode
- Seven-segment LED display
- Common cathode
- Common anode

Practice Questions

1. Give two uses of a switch in a circuit.

   a) ....

   b) ....

2. Draw the circuit symbol for each of the following switches in the boxes below:

   a) Push to Make Switch

   b) Push to Break Switch

3. Which battery has the longest life? Tick the correct option.

   A Zinc carbon ⬭

   B Alkaline ⬭

4. a) Name the four colour bands for a 68K fixed resistor. One has been done for you.

   i) .... ii) .... iii) .... iv) Gold

   b) What does the gold band indicate?

   ....

5. Calculate the value of a current limiting resistor to protect an LED when the battery voltage is 9V and the LED current is 20mA. Show your workings and units.

   Formula: ....

   Working: ....

   ....

   Answer with units: ....

6. Draw the circuit symbol for the following components.

   a) Variable resistor

   b) Potentiometer

   c) Thermistor

   d) Light Dependent Resistor

7 Draw each of the following electrolytic capacitors in the boxes below.

a) Radial capacitor

b) Axial capacitor

8 What is meant by a 'time constant' when referring to capacitors?

9 How many time constants are required to fully charge a capacitor?

10 Calculate the time constant for a capacitor when C = 100 μF and R = 100K. Show your workings and give the units.

Formula:

Working:

Answer with units:

11 Complete the following table. The microfarad has been done for you.

1 microfarad (1uF)	=	10^{-6}F (0.000001F)
1 nanofarad (1nF)	=	**a)**
1 picofarad (1pF)	=	**b)**

12 Circle the correct option in the following sentence.
A forward biased diode **allows current to flow** / **stops current flowing**.

13 Give an example of when you would use a diode in:

a) Forward bias:

b) Reverse bias:

Transistors

Transistors

Transistors are **electronic switches** and **current amplifiers**. There are two types:

- **Bipolar Transistors**.
- **Field Effect Transistors**.

Transistors are packaged in a variety of cases.

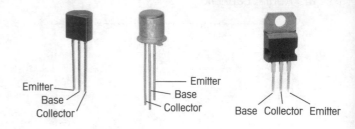

Emitter
Base
Collector

Emitter
Base
Collector

Base Collector Emitter

Bipolar Transistors

Bipolar transistors are divided into…

- **npn** type transistors
- **pnp** type transistors.

npn and pnp refer to the layers of **negative** and **positive semiconductor material**. The only difference between them is the direction of current flow through the transistors. npn transistors are more widely used than pnp transistors.

The semiconductor layers form the leads of the bipolar transistor, which are called…

- the **base** (b)
- the **collector** (c)
- the **emitter** (e).

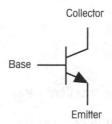

Circuit Symbol For npn Transistor

Collector

Base

Emitter

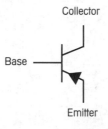

Circuit Symbol For pnp Transistor

Collector

Base

Emitter

Controlling Bipolar Transistors

A bipolar transistor is controlled by applying a voltage of more than 0.7V to the base. A small current flows into the base and out of the emitter. This causes the resistance between the collector and the emitter to fall.

When a transistor's switched off, the resistance between the collector and the emitter is **high**, and when it's switched on the resistance is **low**.

Transistors are **analogue devices** that allow a larger current to flow from the collector to the emitter as the base current increases. A voltage of about 1.5V between the base and the emitter turns the transistor fully on.

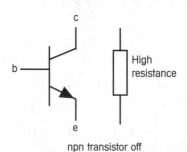

c

b

e

High resistance

npn transistor off

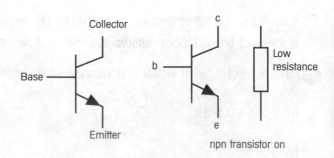

Collector

Base

Emitter

c

b

e

Low resistance

npn transistor on

Transistor Calculations

Two current paths flow through a transistor:

- The base current (small).
- The collector current (large).

Both currents leave the transistor by the emitter lead and become the emitter current. The emitter current (I_e) is the sum of the base current (I_b) and the collector current (I_c).

The **gain** of a transistor is found by dividing the collector current by the base current, represented by the symbol ^hFE.

Formula for calculating the gain:

$$^hFE = \frac{I_c}{I_b}$$

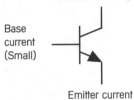

Collector current (large)

Base current (Small)

Emitter current

Formula for calculating the emitter current:

$$I_e = I_b + I_c$$

Transistor Calculation Examples

1 Calculate the gain of a BC548 transistor if the collector current is 100mA when the base current is 0.5mA.

Using the formula:

$$^hFE = \frac{I_c}{I_b}$$

$$= \frac{100\text{mA}}{0.5\text{mA}}$$

$$= \textbf{200}$$

2 The hFE gain of a BC639 transistor is 40 when the collector current is 150mA. Calculate the base current.

$$I_b = \frac{I_c}{^hFE}$$

$$I_b = \frac{150\text{mA}}{40} = \textbf{3.75mA}$$

Quick Test

1 What are the three leads of a bipolar transistor called?
2 True or false – transistors are electronic switches and amplifiers.
3 True or false – bipolar transistors are analogue devices.
4 Write down the formula for working out the gain of a transistor.
5 True or false – $I_e = I_b + I_c$
6 True or false – the base current is bigger than the collector current.

KEY WORDS

Make sure you understand these words before moving on!

- Bipolar transistor
- Field effect transistor
- Base
- Collector
- Emitter
- Gain

Darlington Pair Transistors

Single Transistor Circuit

The **BC548 transistor** has a gain of 220 and a maximum I_c current of 100mA, which is well above the 20mA required by the LED. The base current will therefore be:

$$\frac{\text{Collector current (20mA)}}{\text{Gain (220)}} = 0.09\text{mA}$$

The **input sensor** and the 100K potentiometer form a **potential divider**. When the input sensor's resistance decreases sufficiently, the voltage between the base and emitter increases until it reaches 0.7V and the transistor switches on.

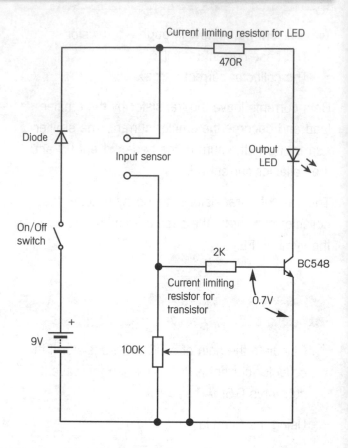

Darlington Pair

The performance of the circuit is improved by adding a second transistor. The second transistor benefits from the gain of the first transistor and the circuit becomes more sensitive to changes at the input.

The second transistor, the **BC639,** also provides a **larger maximum collector current** of 1A, so it can drive electronic devices that require higher currents.

The total gain of a Darlington Pair is the gain (220) of the BC548 transistor multiplied by the gain (40) of the BC639 transistor, which equals 8800.

A typical Darlington transistor used in schools is the **BCX38**, which contains **two transistors in one transistor case**.

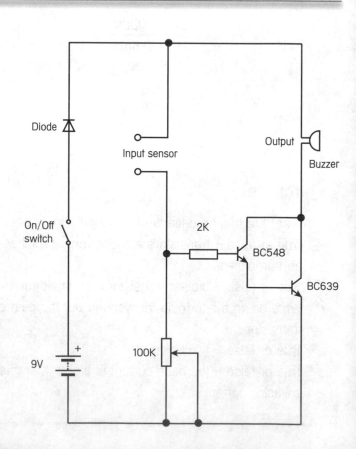

Field Effect Transistors

Field Effect Transistors (FETs) have three leads:
- The **drain** (d).
- The **gate** (g).
- The **source** (s).

A FET amplifies the voltage at its gate to gain an increase in voltage or current. Unlike bipolar transistors, the current flowing between the drain and the source doesn't depend on the size of the current on the gate.

FETs are **voltage controlled** and are examples of a **digital switching action**. Once the gate is triggered by an input voltage…
- above 2V, the FET turns fully on
- below 2V, the FET turns fully off.

Their digital switching action makes them useful as amplifiers for low-powered process units such as…
- **CMOS logic gates**
- **Peripheral Interface Controller**s (PICs)
- electronic switches controlled by **high-resistance inputs**, e.g. touch switches.

FETs are often used to control high currents or to switch high-current devices on and off, e.g. motors.

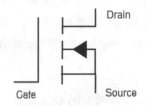

Field Effect Transistor Symbol

Interfacing and Touch Switches

The **interfacing circuit** shows a logic IC or PIC interfaced with a FET. Once the gate's been triggered by an input above 2V, the FET switches on, which allows current to flow to drive the motor.

An advantage of a FET is that very little electrical current is drawn into the gate compared to the base of a bipolar transistor. FETs are voltage controlled and once the gate's been triggered by an input above 2V, the FET turns fully on. Bridging the touch contacts with a finger increases the voltage at the gate and turns on the motor via the FET.

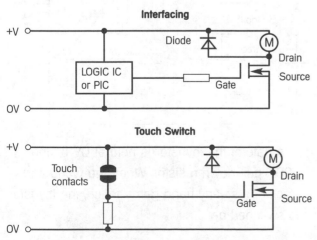

Quick Test

1. What is the gain of a BC548 transistor?
2. What is the gain when a BC548 and a BC639 transistor are connected together as a Darlington pair?
3. Name a typical Darlington transistor used in schools.
4. What does FET stand for?

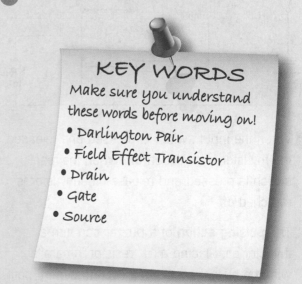

KEY WORDS
Make sure you understand these words before moving on!
- Darlington Pair
- Field Effect Transistor
- Drain
- Gate
- Source

Thyristors

Thyristors

A **thyristor** is like a diode. It has an anode, a cathode, and a third lead called a **gate**. It's used in electronic circuits to control current flow and provide a means of keeping a circuit switched on (latching). A thyristor only allows current to flow from the anode to the cathode when a voltage of more than 2V is applied to the gate.

A thyristor is a **bistable** device (it has two stable states and can only be changed by setting and resetting). A thyristor stays switched on (**latched**) until it's reset by interrupting the current flow through the thyristor, or switching off the power supply.

There's **high resistance** between the anode and cathode leads when a thyristor is switched off. The switching voltage of above 2V causes the resistance between the **anode** and **cathode** to fall and allows current to flow.

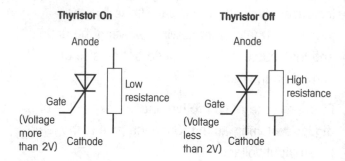

Using Thyristors in Circuits

Thyristors are easy to use and can be switched on by a variety of inputs, for example...

- **tilt** switches
- **vibration** switches
- **membrane** switches
- **reed** switches
- **Piezo transducers**.

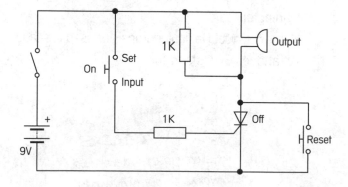

1 When the input switch is pressed and released, the thyristor is switched **on**. When the reset switch is pressed and released the thyristor is switched **off**.

The **pulsing action** of a buzzer can turn a thyristor off. Placing a 1K resistor in parallel with the buzzer provides a second current path.

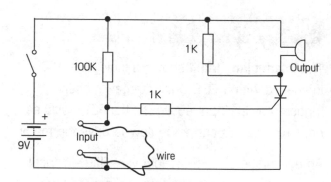

2 The gate of the thyristor is held at 0V by the input wire. When the input is broken, the gate is taken above 2V and the thyristor switches **on**.

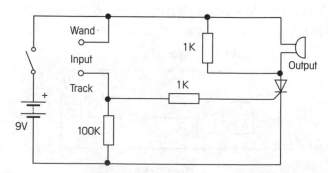

3 The gate of the thyristor is held at 0V by the 100K pull-down resistor. When the wand and track of a steady hand game touch, the thyristor is switched **on**.

Relays

Relays use electromagnetism to enable one electrical circuit to switch on a second circuit without an electrical connection between the two circuits.

When a relay switches off, the **magnetic field** around the coil collapses. This causes a large **voltage spike** to be created in the relay coil. This is 'back emf', which can damage transistors and ICs. A clamping or flywheel diode is used to eliminate this.

Relays are used in many products including cars and household appliances. As well as switching larger currents, a relay can be used to latch a switching circuit.

Disadvantages of relays are that they...
- have slow switching speeds
- are relatively expensive
- can be bulky.

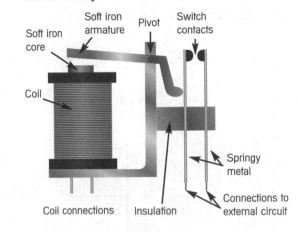

Solenoids

A solenoid is an electromagnetic device consisting of...
- a **coil of wire** wrapped around a **plastic tube**
- a **metal plunger** that fits inside the tube.

When an electrical current is passed through the coil, it becomes magnetic and pulls the plunger into the tube. The plunger can only be made to move **into** the tube – reversing the current flow will not make the plunger move out of the tube. Fitting a spring to the plunger is a way of making the plunger move out.

Solenoids are used to operate **valves** to control the flow of liquids. They can also be used as the locking system of a safe.

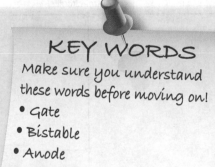

Quick Test

1. What are the three leads of a thyristor called?
2. True or false – a thyristor is a bistable device.
3. What kind of devices are relays and solenoids?
4. Give two examples of where a relay could be used.
5. Name three parts of a relay.

KEY WORDS
Make sure you understand these words before moving on!
- Gate
- Bistable
- Anode
- Cathode
- Back emf
- Electromagnetic

Relay Interfacing

Single Pole Double Throw Relay

The **Single Pole Double Throw (SPDT)** relay can be used to interface a 5V control circuit to a 24V output circuit. The relay allows a low-current control circuit to switch outputs on and off in a **secondary circuit**, which can provide greater power.

When the relay switches off, a large **back emf** is created by the magnetic field collapsing around the coil of the relay. Connecting a reverse biased **clamping diode** across the coil of the relay protects the switching transistor from damage.

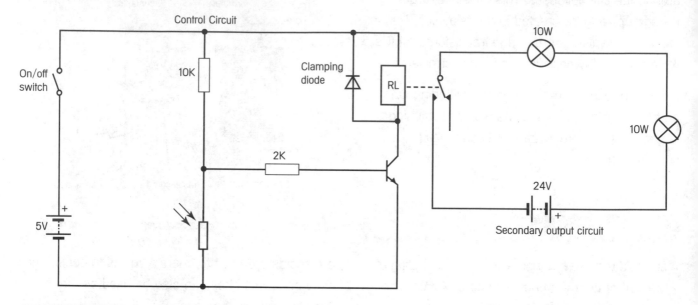

Double Pole Double Throw Relay

The **Double Pole Double Throw (DPDT)** relay is used to provide the same interface as the SPST with the addition of a **latch**. The extra pole and contacts provide an electrical connection across the transistor, connecting the collector to the emitter.

When the transistor switches off, current continues to flow through the relay coil and bypass the transistor through the latch, as shown on the diagram.

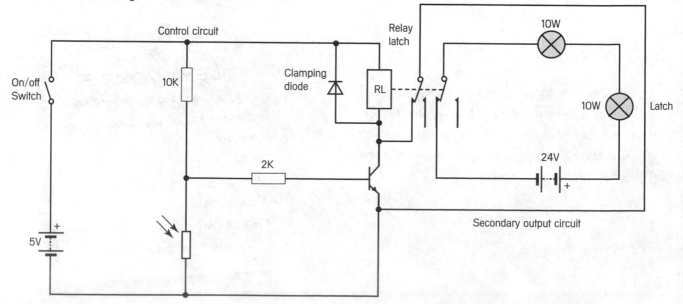

Low Voltage d.c. Motors

Small **d.c. motors** (direct current motors) are often called **toy motors** and are cheap to buy.

An advantage of d.c. motors is that they can be controlled to move in a forward or backward direction, by **reversing** the current flow through the coil of the motor.

A disadvantage is that they have very little accurate control unless they're used with **sensors** to detect distance travelled. Connecting a d.c. motor to a gearbox provides greater control of speed, direction and torque.

Stepper motors…

- are a **special type** of d.c. motor that can be driven in either forward or reverse direction by small accurate steps
- require a power supply that is **continuously pulsed** in specific patterns.

A **PIC microcontroller** can be used to provide the pattern of pulses and a stepper motor driver IC is used to drive the stepper motor. Each pulse to a stepper motor is equal to 7.5 degrees of spindle movement, or 1 step.

A stepper motor completes one revolution in 48 steps. Stepper motors with a 1.8 degree step angle of spindle rotation are also available.

Stepper motors are used in computer printers and are larger and more expensive than standard d.c. motors.

Solar motors are…

- **precision manufactured** d.c. motors of high quality, designed for **low power use**, including solar cells
- expensive when compared with standard d.c. motors.

d.c. Motor Circuits

Electrical noise created by cheap d.c. motors can damage PIC microcontrollers. It can be reduced by soldering a 220nF polyester non-polarised capacitor to the motor contacts.

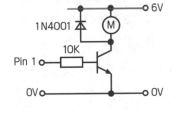

Quick Test

1. When using relays, what is the purpose of a clamping diode?
2. True or false – a relay can be used to latch a circuit.
3. Name three types of low voltage d.c. motors.
4. Name a d.c. motor that gives precise positional control.
5. How can a d.c. motor be controlled to move forwards and backwards?
6. How can electrical noise created by a cheap d.c. motor be reduced?

KEY WORDS

Make sure you understand these words before moving on!
- Clamping diode
- Toy motor
- Electrical noise
- Stepper motor
- Solar motor

Darlington Driver ICs

Darlington Driver ICs

The **ULN2003 IC** contains seven Darlington Driver transistors, similar in value to the BCX38 and **back emf suppression diodes**, so clamping diodes aren't needed.

Pins 1 and 2 are microcontroller outputs.

If it's necessary to pass relatively high currents through an output device, it can be used to pair up drivers, as shown on the diagrams. The ULN2803 IC contains **eight transistors**.

Pins 1 and 2 are microcontroller outputs.

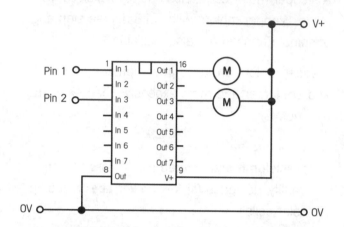

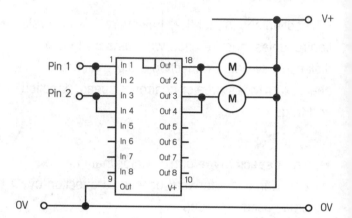

Motor Driver ICs

The **16-pin** four output motor driver L293D IC allows control of a single motor or two motors. Pin 8 of the motor driver IC should be connected to +V. The maximum current is 1A per output.

Pins 4, 5, 6, and 7 are microcontroller outputs.

Changing the states of the microcontroller pins 4 and 5 from high to low will change the direction of current flow through motor A and result in a change of direction of the motor.

Key
Pin 4 and Pin 5 LOW - Motor A Stop
Pin 4 HIGH and Pin 5 LOW - Motor A Forward
Pin 4 LOW and Pin 5 HIGH - Motor A Reverse
Pin 4 and Pin 5 HIGH - Motor A Stop

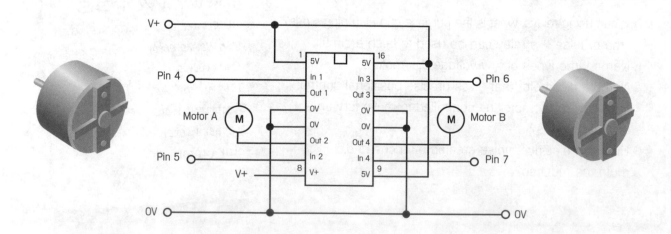

Sounders

Buzzers

Buzzers are **polar components** that change electrical energy into sound. Modern electronic buzzers have built-in amplifier drive circuits and are available with flying leads or for PCB mounting.

Piezo Transducers, Sounders and Buzzers

Piezo transducers…

- change **electrical energy** into **mechanical energy** and vice versa
- can be made very small and are often used as tiny speakers in electronic devices
- can be used to produce many different sounds, unlike buzzers, which only produce a single tone.

The simplest Piezo transducer is the uncased brass disc type with the Piezo material on one side. The material is made from **minerals**, **ceramics** and **polymers**, and demonstrates the **Piezo electric effect** found naturally in quartz.

Piezo sounders and **Piezo buzzers** contain a Piezo transducer and an **electronic oscillator drive circuit**. The drive circuit supplies a varying voltage to the Piezo electric disk. The disk flexes back and forth due to the varying voltage and creates sound. These devices can be quite loud and high-pitched, and make good warning systems.

Loudspeakers

Loudspeakers are devices that convert electrical energy into mechanical energy in the form of **physical vibrations** to create **sound waves** in the air.

At the front of a loudspeaker is a flexible cone:

- The outer part of the cone is fastened to the outer part of the loudspeaker's metal rim.
- The inner part of the cone is fixed to a coil that sits in front of a permanent magnet.

When electrical current is passed through the coil, it turns into an **electromagnet**. As current flows through the coil, it either **attracts** or **repels** the permanent magnet. This moves the coil back and forth, pulling and pushing the loudspeaker cone.

Quick Test

1. What are Darlington driver ICs used for?
2. Name three devices that can be used in a circuit to produce sound.
3. True or false – a buzzer can be connected any way round in a circuit.
4. True or false – a buzzer can only produce a single tone.
5. True or false – a Piezo transducer changes electrical energy into mechanical energy and vice versa.

KEY WORDS

Make sure you understand these words before moving on!
- Darlington driver
- Buzzer
- Piezo transducer
- Piezo sounder
- Piezo buzzer
- Loudspeaker

Practice Questions

1 Choose the correct word from the options given to complete the sentences below.

analogue   **digital**

a) Bipolar transistors are _____ devices.

b) Field Effect Transistors are _____ devices.

2 Draw and label the circuit symbol for each of the transistors below:

a) npn Transistor

b) Field Effect Transistor

3 Which two transistors, when connected together, make a Darlington pair?

a) _____   **b)** _____

4 Fill in the missing words to complete the following sentences.

a) The leads of a transistor are called the base, _____ and emitter.

b) Transistors can be _____ switches.

c) Transistors are made from _____ material.

d) A bipolar transistor is switched on by applying a voltage of _____ V between the base and the emitter.

e) A bipolar transistor allows a large current to flow from the _____ to the _____ as the base current increases.

f) A bipolar transistor is fully switched on if there is a voltage of _____ V between the _____ and emitter.

g) Bipolar transistors are _____ amplifiers.

h) FET transistors are turned off when the gate voltage is less than _____ V.

i) There is a high resistance between the _____ and emitter when a transistor is switched off.

5 Calculate the gain of a transistor if the collector current is 200mA and the base current is 0.5mA. Show your workings and give the units.

Formula: ....

Working: ....

....

Answer with units: ....

6 Calculate the emitter current of a transistor when the collector current is 400mA and the base current is 0.5mA. Show your workings and give the answer with units.

Formula: ....

Working: ....

....

Answer with units: ....

7 Give two ways in which a thyristor can be used in an electronic circuit.

a) ....

b) ....

8 The pulsing action of a buzzer can sometimes turn a thyristor off. How can this be prevented?

....

....

9 What are relays used for? Tick the correct options.

    A Power supply

    B Latching

    C To connect two circuits together with an electrical connection

    D To connect two circuits together without an electrical connection

10 Why are Logic ICs and PIC microcontrollers often interfaced with FETs in electronic circuits?

....

....

555 IC

Integrated Circuits

Integrated Circuits (ICs) have revolutionised electronics, computing and communications, and made a vast range of products possible.

An IC is a **complete miniature electronic circuit** on a **chip**, packaged in a plastic case. The chip is a wafer-thin piece of **silicon**, no more than 5mm square, which can contain thousands of electronic components.

Most ICs are available in rectangular plastic packages with two rows of metal connecting pins. This type of IC package is known as **DIL** (**Dual In Line**).

555 ICs

555 ICs...

- are the main component in **monostable** and **astable** circuits
- can be used in a **monostable circuit** to switch something on for a **certain amount of time** and then off, for example, a courtesy light in a car
- can be used in an **astable circuit** to switch something on and off **continuously,** for example, a flashing warning sign on a motorway.

On top of the IC at one end is a small notch that identifies pin 1 and the highest numbered pin. With the small notch at the top of the IC, pin 1 is always the top left pin and the highest numbered pin is directly opposite. Some ICs also have a small dot to indicate pin 1.

ICs are always numbered in an **anti-clockwise** direction and can have between 4 and 64 pins.

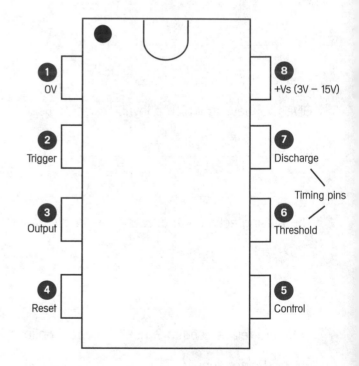

IC Sockets

When using ICs it is good practice to assemble them on the **printed circuit board** in an **IC socket**.

Using an IC socket means there's little chance of the IC being damaged by heat from the **soldering iron** and, if needs be, the IC can be quickly removed without needing to be **de-soldered**.

IC Socket

Letts
and
LONSDALE

ESSENTIALS

GCSE Design & Technology
Electronic Products
Controlled Assessment Guide

About this Guide

The new GCSE Design & Technology courses are assessed through…
- written exam papers
- controlled assessment.

This guide provides…
- an overview of how your course is assessed
- an explanation of controlled assessment
- advice on how best to demonstrate your knowledge and skills in the controlled assessment.

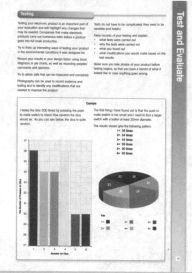

What is Controlled Assessment?

Controlled assessment has replaced coursework. It involves completing a 'design and make' task (two separate tasks for OCR) within a set number of hours.

Your exam board will provide you with a range of tasks to choose from. The purpose of the task(s) is to see how well you can bring all your skills and knowledge together to design and make an original product.

You must produce individual work under controlled conditions, i.e. under the supervision of a teacher.

Your teacher can review your work and give you general feedback. However, all the work must be your own.

How is Controlled Assessment Marked?

Your teacher will mark your work using guidelines from the exam board. A moderator at the exam board will review these marks to ensure that they are fair.

You will not just be marked on the quality of your end product - the other stages of design and development are just as important, if not more so!

This means it is essential to clearly communicate what you did, how you did it, and why you did it, at each stage of the task(s). You will be marked on the quality of your communication too.

This guide looks at the main stages you will need to go through in your controlled assessment task(s), providing helpful advice and tips along the way.

Design

Make

Exam Board	Course	Written Paper	Controlled Assessment
AQA	**Electronic Products** Full Course	2 hours; 120 marks; 40% of total marks Section A (30 marks) – A design question based on a context which you will be notified of before the exam Section B (90 marks) – Covers all the content on the specification, i.e. all the material covered in your Essentials Revision Guide.	Approx. 45 hours; 90 marks; 60% of total marks Consists of a single design and make activity from a range of tasks set by the board.
AQA	**Systems and Control Technology** Full Course	2 hours; 120 marks; 40% of total marks Section A (30 marks) – A design question based on a context which you will be notified of before the exam Section B (90 marks) – Covers all the content on the specification, i.e. all the material covered in your Essentials Revision Guide.	Approx. 45 hours; 90 marks; 60% of total marks Consists of a single design and make activity from a range of tasks set by the board.
Edexcel	**Electronic Products** Full Course	1 hour 30 minutes; 80 marks; 40% of total marks	Approx. 40 hours; 100 marks; 60% of total marks. The design and make activities can be linked (combined design and make) or separate (design one product, make another).
OCR	**Control Systems** Short Course and Full Course	**Sustainable Design** 1 hour; 60 marks; 20% of total marks Section A – 15 multiple choice questions Section B – 3 questions requiring short written answers, sketches and annotations	**Introduction to Designing & Making** 20 hours; 60 marks; 30% of total marks
OCR	**Control Systems** Full Course	**Technical Aspects of Designing and Making** 1 hour 15 minutes; 60 marks; 20% of total marks Section A – 3 questions based on the technical aspects of working with materials, tools and equipment. Section B – 2 questions on the design of products reflecting the wider aspects of sustainability and human use. One of these questions will require a design response.	**Making Quality Products** 20 hours; 60 marks; 30% of total marks

Important Considerations

Unlike your teacher, the moderator will not have the opportunity to see how you progress with the task. They will not be able to talk to you or ask questions – they must make their assessment based on the evidence you provide. That means it is essential to communicate your thoughts, ideas and decisions clearly at each stage of the process:

- Organise your folder so the work is in a logical order.
- Ensure that text is legible and that spelling, punctuation and grammar are accurate.
- Use an appropriate form and style of writing.
- Make sure you use technical terms correctly.

Marks are awarded for quality of presentation but there are more marks available for the content, so make good use of your time – don't waste time creating elaborate borders and titles!

Remember, in Electronic Products your knowledge and understanding of electronics is the most important thing and there should be evidence of this throughout your design folder. You should spend less time on the design and development of the casing or packaging. As a general guide, you should divide your time between electronics and packaging in the ratio of 70:30 or 60:40.

Because you only have a limited amount of time, it is essential to plan ahead. Suggested times for each of the stages are shown. These are guidelines only and will vary depending on the total amount of time your exam board allows for the task(s) (see p.2).

Stage	Tasks	Time
Investigate	Analysing the brief	1 hour
	Research	2-3 hours
	Design specification	1 hour
Design	Initial ideas	5-6 hours
	Review	5-6 hours
Develop	Development	5-6 hours
	Product specification	1-2 hours
Plan	Production Plan	1-2 hours
Make	Manufacture	16 hours
Test and evaluate	Testing and evaluation	1-2 hours

A time plan like this will ensure that you spend a majority of your time working on the areas that are worth the most marks.

That doesn't mean that the other areas aren't important, but quality rather than quantity is the key.

For most projects, 20–30 sheets of A4, or 15–20 sheets of A3, should be sufficient.

A4 size paper is the most popular with students taking Electronic Products.

Analysing the Task

To get the most marks, you need to...
- analyse the task / brief in detail
- clearly identify all the design needs.

It is a good idea to start by writing out the task / brief...
- as it is written by the exam board
- in your own words (to make sure you understand what you are being asked to do).

You then need to identify any specific issues that must be considered before you can start designing the product.

Ask yourself the following questions:
- Who will use the electronic product?
- What will it be used for?
- Where will it be used?
- Are there any cost restrictions that will influence my design?
- Are there any other constraints that I need to consider?
- How many products would be made if it went into commercial production?

Make sure you have clear answers to all these questions before you go any further.

You do not need to write an essay. You could use...
- an attribute analysis table
- a spider diagram
- a list of bullet points.

When analysing a brief for an electronic product, it immediately divides into two parts:
- the electronics
- the case (that holds the electronic circuitry and input and output devices).

Shown below are some of the things you might consider.

You should spend more time on the electronics than the case.

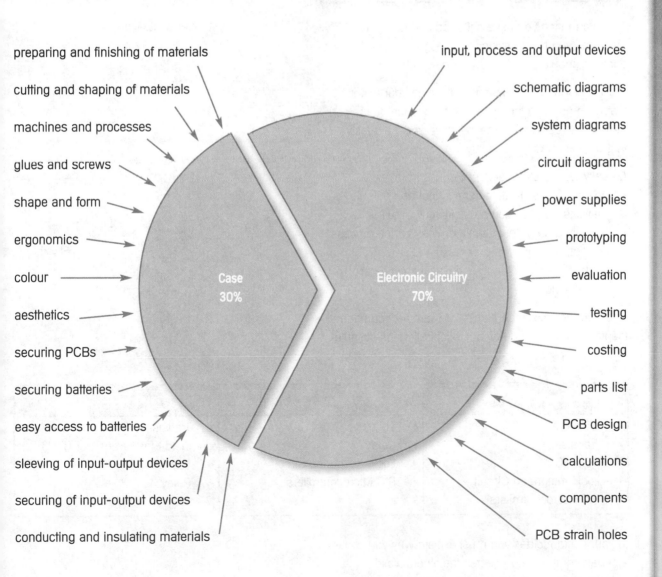

preparing and finishing of materials

cutting and shaping of materials

machines and processes

glues and screws

shape and form

ergonomics

colour

aesthetics

securing PCBs

securing batteries

easy access to batteries

sleeving of input-output devices

securing of input-output devices

conducting and insulating materials

Case 30%

Electronic Circuitry 70%

input, process and output devices

schematic diagrams

system diagrams

circuit diagrams

power supplies

prototyping

evaluation

testing

costing

parts list

PCB design

calculations

components

PCB strain holes

Product Analysis

Product analysis is a useful way of identifying what research you need to carry out. It involves looking carefully at an existing product and investigating how it works. It is especially useful for identifying…
- input devices
- output devices
- how the PCB and battery are fitted into the case.

Unfortunately, product analysis will not show how the process section of an electronic product works, i.e. how the electronic components and ICs work together. You must use your imagination and decide what takes place in the process section by observing and recording the input(s) and output(s).

You can use this information to develop a systems analysis for the product.

A Standard Smoke Alarm

The Inside of a Smoke Alarm

Research

The information you have collected and the systems diagrams you have developed through product analysis should help focus your research.

Do not include any research material in your folder that does not help the development of your ideas. Remember, designing time is limited and it's quality that earns marks not quantity.

An easy way to start your research is to…
- identify two or three different input and output devices that you could use in your circuit
- identify a range of suitable devices for the process section (a slightly harder task!)
- look at the advantages and disadvantages of each device.

If you are designing an alarm, for example, you might identify Logic ICs and PIC microcontrollers as suitable process blocks:

Basic Block Systems Diagram

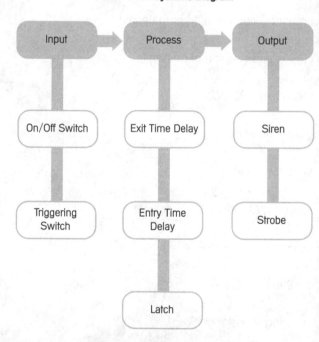

To make a monostable time delay	To make a latch	As a transducer driver
• 555 IC • Transistor • 4047 IC • Logic Integrated Circuit • PIC Microcontrollers	• Thyristor • Relay • Logic Integrated Circuit • PIC Microcontrollers	• Transistor • Field Effect Transistor (FET) • Thyristor • Relay

In your design folder, you must explain why you chose a certain method and rejected the others. Valid reasons might relate to cost, size, the working parameters of components or availability.

Interviews, Questionnaires and Surveys

Before you spend time on interviews, questionnaires and surveys, ask yourself why you are doing it. If you cannot come up with a good answer, then don't do it. If you feel the information you will gain from these activities will help your design thinking, then go ahead, but only assign one page of A4 or A3 to this activity, including the analysis of results!

Great skill is required to produce a good questionnaire. The questions should be short and not open-ended. Ideally, they should be answerable by ticking a range of boxes. This enables the results to be analysed easily. Questionnaires can be photocopied and handed out.

Remember, asking ten people is unlikely to provide you with a valid survey result. You need to target potential end-users and obtain a good cross-section of opinions if your data is to be reliable.

Try to think of ten short, quality questions. Once you have collated the results, you can produce line graphs, bar charts and pie charts.

Data bases or spreadsheets can help you collate the results of your questionnaire and can generate useful graphs. It is essential to explain how the survey was carried out, what questions were asked and how the results influenced your thinking.

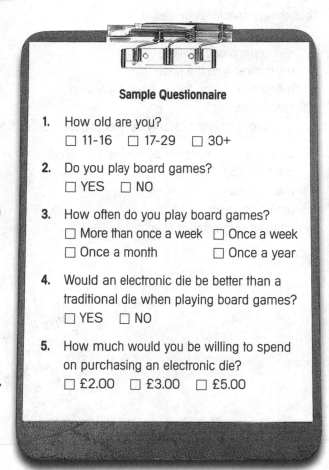

Sample Questionnaire

1. How old are you?
   ☐ 11-16   ☐ 17-29   ☐ 30+

2. Do you play board games?
   ☐ YES   ☐ NO

3. How often do you play board games?
   ☐ More than once a week   ☐ Once a week
   ☐ Once a month   ☐ Once a year

4. Would an electronic die be better than a traditional die when playing board games?
   ☐ YES   ☐ NO

5. How much would you be willing to spend on purchasing an electronic die?
   ☐ £2.00   ☐ £3.00   ☐ £5.00

Line Graphs	Bar Charts	Pie Charts
These are useful for showing changes over time. They are a set of dots (or crosses), with adjacent dots joined together by a straight line.	These are useful for making comparisons of results.	These are useful for representing information as a percentage to show clear comparisons.

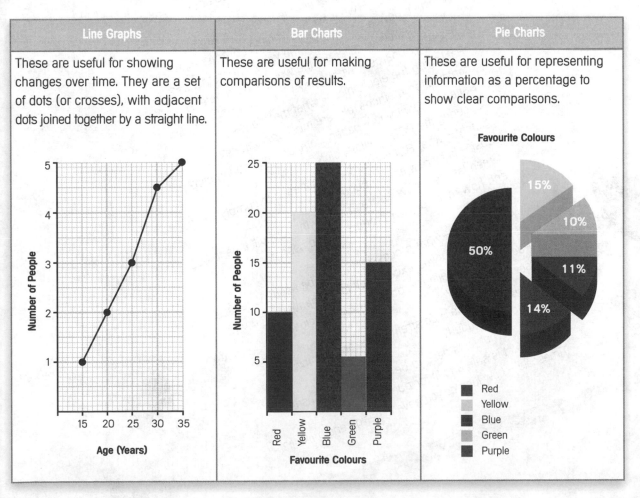

Design Specification

Your design specification should…
- provide a detailed description of what the electronic product will be
- relate directly to the brief
- reflect information found in your research and analysis
- be clearly presented (usually as a list of bullet points).

It is essential that the points on your specification…
- are realistic / achievable
- are technical
- are measurable
- address some issues of sustainability.

A good specification is crucial to the success of any Electronic Products project and will make it easier for you to carry out on-going and final evaluation. It may be that your specification is re-written a number of times as you proceed with designing.

Your design specification should include some of the following :

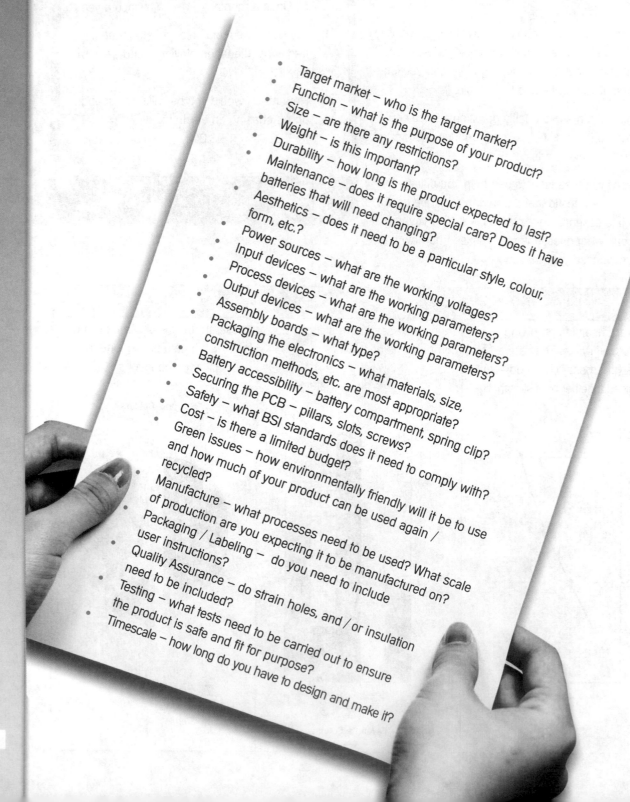

- Target market – who is the target market?
- Function – what is the purpose of your product?
- Size – are there any restrictions?
- Weight – is this important?
- Durability – how long is the product expected to last?
- Maintenance – does it require special care? Does it have batteries that will need changing?
- Aesthetics – does it need to be a particular style, colour, form, etc.?
- Power sources – what are the working voltages?
- Input devices – what are the working parameters?
- Process devices – what are the working parameters?
- Output devices – what are the working parameters?
- Assembly boards – what type?
- Packaging the electronics – what materials, size, construction methods, etc. are most appropriate?
- Battery accessibility – battery compartment, spring clip?
- Securing the PCB – pillars, slots, screws?
- Safety – what BSI standards does it need to comply with?
- Cost – is there a limited budget?
- Green issues – how environmentally friendly will it be to use and how much of your product can be used again / recycled?
- Manufacture – what processes need to be used? What scale of production are you expecting it to be manufactured on?
- Packaging / Labeling – do you need to include user instructions?
- Quality Assurance – do strain holes, and / or insulation need to be included?
- Testing – what tests need to be carried out to ensure the product is safe and fit for purpose?
- Timescale – how long do you have to design and make it?

Initial Ideas

Initial ideas are an important part of any design process, and you should allow yourself plenty of time for this stage.

This is your chance to show off your designing skills, but make sure your ideas...

- include three circuit ideas and two case ideas
- are clearly presented
- are realistic and workable
- address all the essential criteria on your design specification
- are presented as drawings and sketches with notes and annotation.

Where possible make good use of CAD.

Circuit Wizard

Logicator for PIC Micros

Electronic Circuit

At GCSE level, you are not expected to design original electronic circuits but to select and modify existing circuits to meet your needs. This may involve...

- joining together two circuits
- changing the input and output devices of a circuit
- finding a latching device
- any other suitable modifications.

Gather and explore circuits from any suitable resource to help you with your Initial Ideas. This can include materials from books, data sheets and computer generated information.

You can sketch or draw your ideas by any means. Where possible try to make good use of CAD and include a variety of schematic, block and circuit diagrams.

You should try to devise tests for your circuits using kits, software packages and breadboards. Record the results using appropriate measuring instruments. Photography provides an excellent record of experimental work and will enhance your design folder.

Try to include mathematical calculations. Work on potential dividers, component ratings, time delays, frequency, current drain, battery life and the size of protective resistors are all examples where calculations can be applied. Including calculations in your controlled assessment is also good preparation for the written paper.

Case or Packaging

Case ideas should be relatively simple and appropriate to house an electronic circuit. A wide range of materials can be used including resistant materials, textiles, card and re-cycled materials.

The drawings for your two ideas should...

- be fully dimensioned
- show positions of input and output devices
- show how the PCB and battery are secured.

Individually designed cases or bought-in cases are acceptable. It is the modifications to the bought-in case that you make to accommodate the electronic circuit that gain the marks. Evidence of design modifications should be shown in design folders as drawings, sketches and notes.

Reviewing Ideas

You need to review your initial ideas to select one or two to develop further. They must satisfy the essential criteria on your design specification, but you will also want to consider...

- which designs satisfy the most desirable criteria
- which designs are most unique / innovative

- which designs are most appealing / attractive
- which designs are cost effective
- which designs are technologically most advanced.

Developing Ideas

Your aim at this stage is to modify and revise your three initial circuit ideas and your two case ideas until you reach the best possible design solution. This will be guided by the initial ideas review you carried out.

When your teacher and the moderator look at your development sheets they will expect to see a design that is improved compared to your initial ideas.

Electronic Circuit

When developing your solution, you will need to give reasons…
* why you have selected a certain circuit and a case design from your initial ideas
* why you have rejected the others designs.

You may have decided to take a number of sub-systems from the three different circuits, in which case you need to explain why you did this.

You should present an accurate final circuit drawing that satisfies the specification and clearly takes into account relevant research and analysis.

The circuit diagram should contain sufficient information for the circuit to be made by a competent third person.

You should then design the component layout. You can choose from a variety of assembly boards, from printed circuit boards to veroboard. Whatever method is used, you will need to show evidence of planning the layout of the circuit for…
* ease of component assembly
* soldering
* inspection purposes
* position of input and output devices
* securing of the circuit board in the external package.

If veroboard is used, for example, you should show recorded evidence in your design folder of planning the component layout, the number of link wires required and the position of breaks in the conductive tracks.

If you intend to use a printed circuit board, you should show the developmental stages of the PCB layout or transparent overlay. When designing the PCB mask, you should try to make the circuit as small as it is practically possible, but make sure…
* the tracks of the PCB are wide enough to withstand the etching process
* the pads are large enough to aid the soldering process.

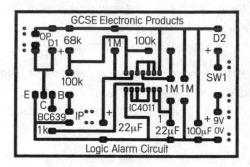

Many poorly soldered circuits are the result of a badly designed PCB.

Case or Packaging

All drawings relating to cases and packaging should be fully dimensioned:
* For vacuum formed cases, include the design of the former showing draft angles and radii on the corners.
* For laser-cut or rapid prototyped cases, include CAD / CAM files of the finished design.
* For bought-in cases, clearly show all the modifications to the case in detail.
* For packaging made from textiles or leather, show evidence of the packaging by drawing sketches or photographs.
* For promotional displays which can be packaged in thick card, include drawings of the flat pack products.

Alarm Case
* Box constructed with butt joints and PVA glue
* Top made from Perspex
* Base made from plywood
* Sides made from pine

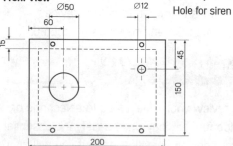

Using ICT

You should use a range of communication techniques and media, including ICT and CAD, where appropriate throughout the design and make task(s). This is particularly important at the development stage.

This can include…
- standard applications, like Word or Excel
- specialist software
- digital camera
- scanner
- plotter / cutter
- CAD / CAM.

Computer Aided Design

Using CAD systems will make it quicker and easier to complete the task and will help you to create a quality electronic product.

If it is available to you and it is appropriate, you should try to use CAD as much as possible.

You might use CAD to…
- make templates
- improve the accuracy and clarity of your drawings
- create numerical data for use on CNC machinery
- test and simulate electronic circuits
- design schematic circuits
- design systems diagrams
- design PCB layouts
- design the case or packaging for your project
- design PIC microcontroller programmes.

Make sure you include a series of screen grabs in your project folder, so the moderator can see how you used CAD, the changes you made, and how it was used to set up any CAM equipment.

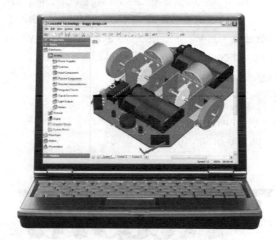

PIC Microcontrollers

If you intend to use a PIC microcontroller as part of your project, you must show evidence in your design folder of the development stages of the PIC programme:
- Use annotation to explain the changes you have made and make sure to include screen grab evidence.

- Try to show an understanding of the relationship between the PIC and the discrete process building blocks which it has replaced.
- When describing a PIC programme, always refer to the ICs input pins and output pins.

The most popular PICAXE Microcontrollers used in schools								
PICAXE Range	IC Size	Memory (lines)	I/O Pins	Outputs	Inputs	ADC	Data Memory	Infra-red
PICAXE-08M	8	80	5	1-4	1-4	3	256 – prog	Yes
PICAXE-14M	14	80	11	6	5	2	256 – prog	Yes
PICAXE-20M	20	80	16	8	8	4	256 – prog	Yes

Production Plan

Your production plan should show...
- the different stages of manufacture in the correct order
- when and what quality control checks will take place.

A flow chart might be the best way of presenting your production plan although sometimes a simple chart listing the stages and the equipment that will be used is equally suitable.

If you draw a flow chart, there are different, specific symbols for each stage of the process. Some are shown opposite.

The symbols are linked together by arrows to show the correct sequence of events.

You should aim to keep your flow chart as clear and simple as possible.

Terminator

represents start, restart, stop

Decision

represents a choice which can lead to another pathway

Process

represents a particular instruction or action

Input/Output

represents additions to or removals from the particular process

Component List

A component parts list is essential when planning to make an electronic product. Always make sure the components you want to use are available and within your project cost. Look in the Rapid Electronics catalogue and become familiar with the data needed to order component parts.

Below is an example of a components list with costings and order numbers.

Quantity	Description	Order Number	Cost
1	Battery Clip (Heavy Duty)	18-0092	5p
1	SPST Rocker Switch (S1)	78-0480	30p
1	Push to Make Switch (S2)	78-1520	20p
2	10K Resistor	62-0397	1p
1	22K Resistor	62-0405	1p
7	330 Resistor	62-0361	1p
1	8 Pin IC Socket	22-0150	3p
7	5mm Red LED	55-0150	5p
1	Battery Holder (3 x AA)	18-0127	15p

Health and Safety

Before you begin making your product, always carry out a risk assessment.

Look at each stage of your production plan in turn and make a list of possible health and safety risks.

Work back through the list and plan how you will minimize the risks at each stage, e.g. by wearing safety equipment, by ensuring you know how to use the tools correctly, etc.

Make sure you include your risk assessment in your flowchart or as part of your planning chart.

Due to the danger of mains electricity, do not use it to power your circuit.

Manufacture

Your revision guide includes information on many of the materials, tools, processes and methods relevant to your particular subject.

In making the prototype of your final product you should demonstrate that, for each specific task, you can select and use...
- appropriate tools and equipment
- appropriate processes, methods and techniques (including CAD / CAM where relevant)
- appropriate materials and components
- all the above correctly and safely.

You should aim to demonstrate a wide range of skills and processes with precision and accuracy.

Remember, all the materials, components, methods and processes that you choose must help to make your product the best possible design solution for the brief. Don't include something just to show off your skills!

The finished product should be...
- an electronic circuit housed in a case or packaged in a suitable material
- accurately assembled
- well finished
- fully functional.

Don't worry if it doesn't turn out quite the way you hoped though - you will earn marks for all the skills and processes you demonstrate, so make sure you record them all clearly in your folder.

If your circuit does not work as planned, make sure that you have carried out and recorded a number of fault finding activities. List the test equipment you used and the readings you took at various points on the electronic circuit.

For each stage of production, make sure you include an explanation of any safety measures you had to take.

You can use photographs to create a pictorial flow chart to show the moderator the making process from start to finish.

Industry

By now you should have a good understanding of the methods and processes used in the design and manufacture of electronic products.

Although you will probably only produce one final product, it is important to show that you are aware of various possible methods of production and how your product would be manufactured commercially. You should explain this in your design folder, for example, compare your one-off production against batch production.

If your product could potentially be manufactured using several different methods, try to list the pros and cons for each method and then use these lists to make a decision about which method you would recommend, for example, compare the through hole method against the surface mount component method.

If you know that a method or process you are using to make your product would be carried out differently in a factory, make a note of this in your project folder, for example, compare assembling components by hand against a pick and place component machine – this will show your teacher and moderator how much you know!

Quality Control and Assurance

Your revision guide looks at some of the quality control tests and quality assurance checks used in industry that are relevant to electronic products.

You may be able to apply some to your own product.

Even if you can't, you should still identify and explain the ones that would be relevant to your product if it were to be produced commercially.

The quality checks you need to make throughout the manufacturing stages should also be included in your flowchart or making plan. It is also important to show any modifications needed to improve the quality of the product.

Logic Probe

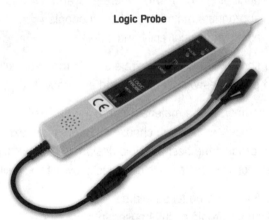

Multimeter

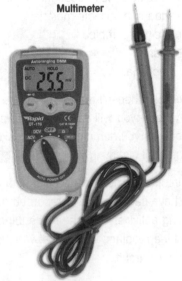

Oscilloscope

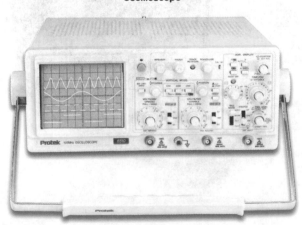

Testing

Testing your electronic product is an important part of your evaluation and will highlight any changes that may be needed. Companies that make electronic products carry out numerous tests before a product goes into full scale production.

Try to think up interesting ways of testing your product in the environmental conditions it was designed for.

Record your results in your design folder using block diagrams or pie charts, as well as recording peoples' comments and opinions.

Try to obtain data that can be measured and compared.

Photography can be used to record evidence and testing and to identify any modifications that are needed to improve the product.

Tests do not have to be complicated; they need to be sensible and helpful.

Keep records of your testing and explain…
- what tests were carried out
- why the tests were carried out
- what you found out
- what modifications you would make based on the test results.

Make sure you take photos of your product before testing begins, so that you have a record of what it looked like in case anything goes wrong.

Example

I tested the dice 200 times by pressing the push to make switch to check how random the dice would be. As you can see below, the dice is quite random.

The first thing I have found out is that the push to make switch is too small and I need to find a larger switch with a button at least 20mm diameter.

The results shown give the following pattern:

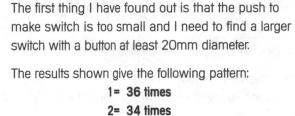

1= **36 times**
2= **34 times**
3= **34 times**
4= **36 times**
5= **30 times**
6= **30 times**

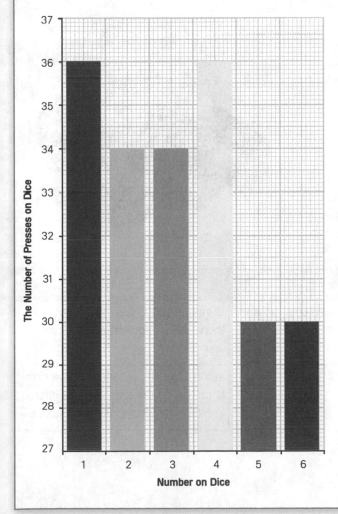

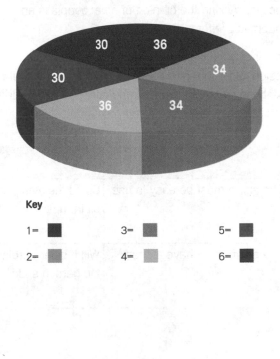

Key

1= █   3= █   5= █
2= █   4= █   6= █

Evaluation

The final evaluation should summarise all your earlier conclusions (from research and reviews) and provide an objective evaluation of the electronic product.

It should refer to the criteria on your original design specification directly.

When carrying out an evaluation, you should ask yourself the following questions:
- Does the product meet all the criteria on the original brief / specification?
- Is the product easy to use?
- Does the product function the way it was intended to?
- What do you think of the style of the product?
- Do you like / dislike any features? Explain why.
- Would you purchase the product and what would you be prepared to pay for it?
- What are its advantages and disadvantages compared to similar products?
- What impact does making and using the product have on the environment?

It is important to ask these questions of end-users too!

Depending on the answers to these questions, you can include suggestions for further modifications in your evaluation.

Honesty is the best policy when writing evaluations. If something didn't work, say so – but always suggest a way of preventing the same problem in future.

As well as using your own judgment you can use the judgments of end-users to assess your product's success. Asking the opinion of other people is an important part of evaluation.

The final evaluation is one of the few places in a design folder that gives you the chance to carry out an extended piece of writing. Remember there are marks available for the quality of your communication.

EVALUATION SUMMARY AGAINST THE SPECIFICATION

1. I spent more than 40 hours designing and making the display. The problem was more complex and demanding than I expected.

2. The display attracts customers' attention by the use of flashing LEDs.

3. The display activates when a person stands in front of it up to a distance of 2 metres.

4. The display is powered by 6 x 1.5 volt cells.

5. Now that I have changed the operational amplifier from a 741 IC to a CA3140E MOSFET Op.Amp., the system will work within a voltage range of 3V to 15V.

6. The PCB is secured.

7. The battery is secured.

8. The battery can easily be replaced.

9. The display has cost less than £10.

10. The display works and is complete.

Specification criteria	Test or question	Results and explanations
The alarm must be easy to use.	Could the alarm be switched on and off in the entry and exit time delay?	During daylight, yes, but in darkness there was a problem and a torch was needed.
The alarm must have a siren.	Will it deter burglars who break into the garden shed?	In the shed, the loudness of the siren is painful on the ears, and the siren can be heard from inside the house.

555 IC Monostable

The 555 IC is in the **monostable state** until pin 2 is triggered by a low voltage.

After the time period it always returns to the monostable state.

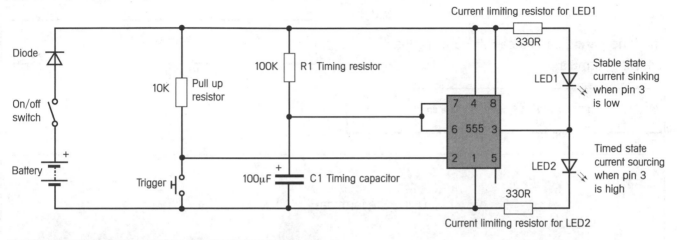

Current limiting resistor for LED1
330R
Diode
On/off switch
Battery
10K Pull up resistor
100K R1 Timing resistor
Trigger
100µF C1 Timing capacitor
7 4 8
6 555 3
2 1 5
LED1 Stable state current sinking when pin 3 is low
LED2 Timed state current sourcing when pin 3 is high
330R
Current limiting resistor for LED2

How the 555 IC Monostable Works

555 IC monostable circuits work in the following way:

1 When the IC is in the **monostable state**, pin 2 is **high** voltage. Pin 3 is **low** voltage with current **sinking** in and LED1 is lit.

2 When pin 2 is triggered by being taken to low voltage by a **switch** or an **electronic signal**, the output at pin 3 changes from 0V to 7V and stays in the timed state for a set amount of time. Pin 3 is now **sourcing** current and LED2 is now lit.

3 The **monostable time period** is set by the **timing resistor** (R1) and the **timing capacitor** (C1), which are connected to timing pins 6 and 7.

4 When the timed period is complete, the output at pin 3 changes from 7V to 0V and stays in this monostable state until the circuit is triggered or switched off.

5 The 10K pull-up resistor connected to the trigger pin 2 ensures that the voltage on the trigger pin is kept **high** until triggered by a **low input**.

6 The 555 IC uses 2 volts internally, so the output is 7V.

Quick Test

1 True or false – 555 ICs can be used as monostable or astable timers.
2 What kind of timer does a flashing warning sign on a motorway use?
3 What kind of timer does a courtesy light in a car use?
4 How is pin 1 indicated on a 555 IC?
5 What pin number is directly opposite pin 1 on a 555 IC?

KEY WORDS
Make sure you understand these words before moving on!
• Monostable
• Astable
• Integrated circuit
• Sinking
• Sourcing
• Monostable time period

555 IC Monostable Calculations

Calculating the 555 IC Monostable Time Period

The length of time that the **monostable** switches on for after pin 2 is triggered is dependent upon the size of the timing resistor (R1) and the timing capacitor (C1).

The **time period** can be calculated using the following formula:

$$T = 1.1 \ X \ CR, \text{ where 1.1 is a constant}$$

Due to electrolytic capacitors having a **large tolerance** of ± **20%**, the formula can be shortened:

$$T = C \ X \ R$$

Example

Using the formula T = 1.1 x CR

T = 1.1 x CR
T = 1.1 x 100μF x 100K
T = 1.1 x 100 x 10^{-6} x 100 x 10^{3}
T = 11 seconds

OR

T = 1.1 x CR
T = 1.1 x 100μF x 100K
$$T = \frac{1.1 \times 100 \times 100 \times 1\,000}{1\,000\,000}$$
T = 11 seconds

Using the formula T = C x R

T = C x R
T = 100μF x 100K
T = 100 x 10^{-6} x 100 x 10^{3}
T = 10 seconds

OR

T = C x R
T = 100μF x 100K
$$T = \frac{100 \times 100 \times 1\,000}{1\,000\,000}$$
T = 10 seconds

Designing Monostable Circuits

If a more accurate time delay is required, a **potentiometer** or **variable resistor** can be included.

When designing monostable circuits…

- the **minimum value** of **R1** should be **1K**
- the **maximum value** of **R1** should be **1M**
- the **minimum value** of **C1** should be **100pF**
- the **maximum value** of **C1** should be **1000μF**
- the **minimum time period** is about **0.1μs**
- the **maximum time period** is about **1000s**.

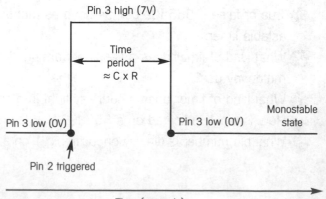

555 IC Astable

When an **astable** circuit is first switched on, the timing capacitor C1 is not charged up. The voltage across the capacitor is less than one-third of the battery voltage. This causes the output voltage at pin 3 to be **high**.

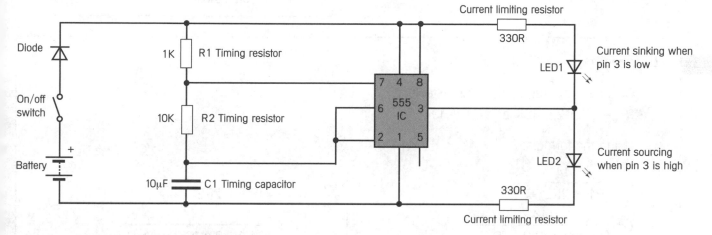

How a 555 IC Astable Circuit Works

555 IC astable circuits work in the following way:

- Capacitor C1 **charges** through the timing resistors R1 and R2 until the voltage across C1 is **greater** than two-thirds of the battery voltage. At this point, the output voltage at pin 3 changes from high voltage to OV.
- Capacitor C1 then **discharges** through R2 into pin 7 until the voltage across C1 becomes less than one-third of the battery voltage.

When this happens, the output at pin 3 changes from OV to high voltage and repeats the process until the circuit is switched off.

When designing astable circuits…

- use **non-electrolytic capacitors** for accurate timing
- for an approximate equal mark–space ratio, make R1 = 1K and R2 = 1M.

Quick Test

1. Write down the formula for calculating the monostable time period for a 555 IC.
2. When a 555 IC is in the stable state, is pin 3 high or low?
3. True or false – when pin 3 on a 555 IC is high, current is sinking into pin 3.
4. On a 555 IC, which pin is the trigger pin?
5. What happens when pin 2 on a 555 IC is taken low?

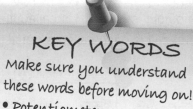

KEY WORDS
Make sure you understand these words before moving on!
- Potentiometer
- Variable resistor
- Astable

555 IC Astable Calculations

555 IC Astable Calculations

The number of **pulses** a 555 IC operating in astable mode makes in one second is called the **frequency**.

The frequency of the 555 IC is determined by the size of resistor 1 (R1), resistor 2 (R2) and the capacitor (C1).

The unit of measurement for frequency is the **hertz** (Hz). The frequency of a 555 IC can be calculated using this formula:

$$f = \frac{1.44}{(R1 + 2R2)C}$$ where 1.44 is a constant

Example

Calculate the frequency of an astable if R1 = 1K, R2 = 10K and C1 = 10μF.

Using the formula:

$$f = \frac{144}{(R1 + 2R2)\ C}$$

$$f = \frac{144}{(1K + 2 \times 10K) \times 10μF}$$

$$f = \frac{144}{(1 \times 10^3 + [2 \times 10 \times 10^3]) \times 10 \times 10^{-6}}$$

$$f = \frac{1.44}{0.21}$$

$$f = 6.9Hz$$

(Which means there are just under 7 pulses every second)

OR

$$f = \frac{144}{(R1 + 2R2)\ C}$$

$$f = \frac{144}{(1K + 2 \times 10K) \times 10μF}$$

$$f = \frac{1.44 \times 1000\,000}{(1\,000 + 20\,000) \times 10}$$

$$f = \frac{1\,440\,000}{210\,000}$$

$$f = 6.9Hz$$

Mark–Space Ratio

The **Mark–Space Ratio** is the length of time the output is high (on) and is called the **Mark**. The length of time the output is low (off) is called the **space**. An astable circuit is often called a **pulse generator**.

The timing capacitor charges through R1 and R2, but only discharges through R2. This makes it difficult to have an equal mark–space ratio.

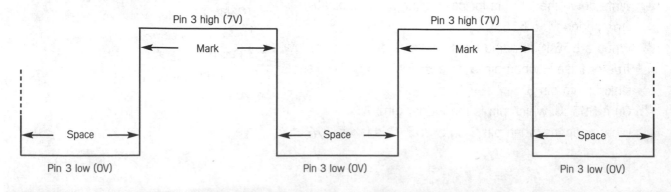

555 IC Monostable & 555 IC Astable

Two 555 ICs

When the PTMS is pressed, LED1 and LED2 flash alternatively at a frequency set by the 555 IC astable for a **time constant** set by the 555 IC monostable.

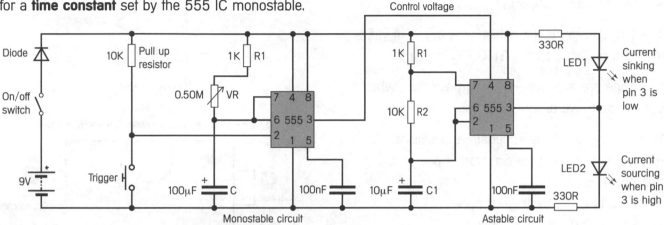

Monostable circuit

Astable circuit

Mark–Space Diagrams

The **mark–space diagram** shows what happens...
- at the **monostable circuit**

- in the **astable circuit**, when the output from the monostable circuit is **high**.

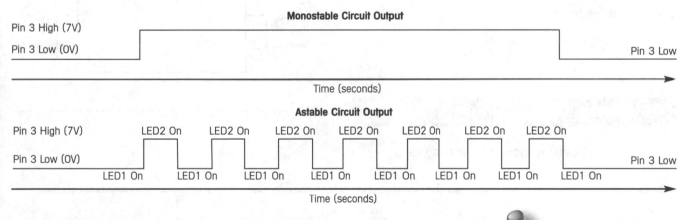

Quick Test

1. Name the unit of measurement for frequency.
2. What is one pulse per second equal to?
3. Which three components control the frequency of a 555 IC working as an astable?
4. True or false – making the capacitor a smaller size increases the frequency of the 555 IC.
5. True or false – the output from a 555 IC is a digital output.

Operational Amplifiers

Operational Amplifiers

Operational amplifiers (op-amps) have two inputs:

- The **inverting input** (marked -).
- The **non-inverting input** (marked +).

The 741 op-amp is the oldest type. In recent years it's been replaced by the 3140 FET op-amp.

Op-amps are available in **DIL IC packages**, which contain one, two or four op-amps.

When using op-amps in certain applications, it's necessary to have a special type of power supply that provides…

- a **positive** (+V) voltage supply
- a zero (OV) voltage supply
- a **negative** (-V) voltage supply.

A **dual power supply** of +9V and -9V can be made by connecting two PP3 batteries together in series. The common connection between the two batteries will be zero volts.

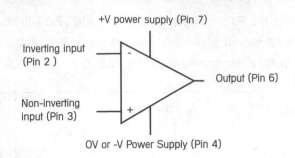

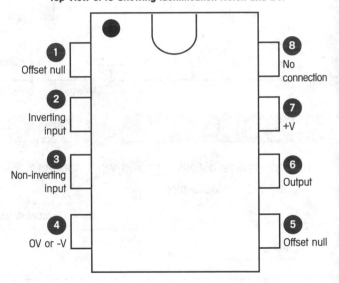

Top View of IC Showing Identification Notch and Dot

Op-Amps with a Dual Power Supply

The diagram shows an op–amp working as a comparator.

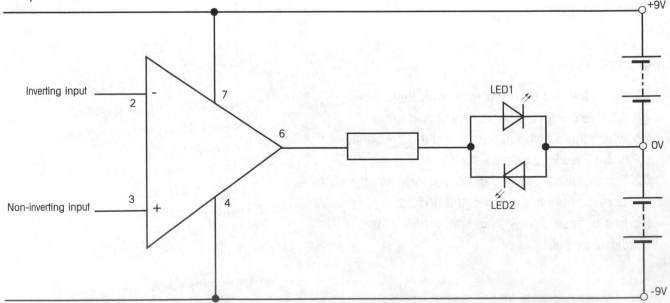

Op-Amp Working as a Comparator

The Op-Amp as a Comparator

If the non-inverting input voltage is **bigger** than the **inverting** input voltage, the output is **high voltage**. If the inverting input voltage is **bigger** than the **non-inverting** input voltage, the output is **low voltage**.

The op-amp detects very small changes in voltage between the two inputs and then multiplies the difference by the gain of the op-amp (around 100 000 on **open loop** gain).

Any slight difference between the input voltages causes the output to **swing** to high or low voltage.

This means that the op-amp is an **analogue-to-digital converter** and is suitable for connecting analogue sensors to logic circuits.

Pin 2	Pin 3	Pin 6
Inverting Input Voltage	Non-inverting Input Voltage	Output Voltage
5V	6V	High
6V	5V	Low
4.5V	4.2V	Low
4.5V	4.6V	High
4.50V	4.51V	High

Output Voltage Swing of an Op-Amp

In practice, 741 op-amps have a **voltage output swing** that is equal to the power supply, minus approximately 2V, which is used internally.

The voltage output swing from a 741 op-amp operating with a...
- 9V dual power supply would be +7V and -7V
- 9V single power supply would be +7V and +2V.

741 op-amps powered by a single power supply only go as low as 2V output.

This can cause problems when trying to use op-amps to switch off transistors and ICs because these components need a **lower voltage**.

This problem can be overcome by using a **3140 FET op-amp**, which operates on a 9V single power supply and outputs to a maximum voltage of 7V and a minimum voltage of 0V.

The 3140 FET op-amp is a direct replacement for the 741, so it can be used in circuits designed for the 741 op-amp.

Quick Test

1. Which two pins are connected to the power rails on an op-amp?
2. How many inputs and outputs does an op-amp have?
3. What is the gain of an op-amp?
4. On an op-amp, which pin is the output pin?
5. How do you make a dual power supply of +9V and -9V?
6. True or false — when pin 2 has a bigger input voltage than pin 3, the output of an op-amp is low.

KEY WORDS
Make sure you understand these words before moving on!
- Operational amplifier
- Inverting input
- Non-inverting input
- Dual power supply

The Op-Amp as an Inverting Amplifier

Using the Op-Amp as an Inverting Amplifier — AQA

With **Open Loop**, the gain of the op-amp is around **100 000** and causes the output of the op-amp to swing from high to low voltage, giving a **digital output**. This output action can cause **distortion** in some op-amp applications by **clipping** the output signal. Clipping occurs when the input signal is amplified beyond the size of the power supply.

As the op-amp has a gain of 100 000, a difference between the pin 2 and pin 3 inputs of 0.001V would need an output of 100V, which is well beyond the supply voltage of, for example, 9V. It's necessary on these occasions to set the gain to a desired level using **negative feedback**.

When an op-amp is used as an **inverting amplifier**, negative feedback is used to reduce the overall voltage gain of the circuit. In the inverting amplifier, the input voltage is applied via the input resistor Rin to the inverting input terminal. The non-inverting input is connected to 0V. Negative feedback is provided by the feedback resistor Rf.

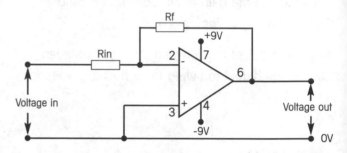

Setting the Voltage Gain — AQA

The gain of an op-amp is set by using a **feedback resistor** (Rf) and an **input resistor** (Rin). The feedback resistor, by feeding back a small part of the output to the inverting input, ensures that the feedback is **negative**.

The advantage of negative feedback is that the op-amp is more **stable** and the gain **predictable**. The formula for calculating the gain of an inverting op-amp is:

$$\text{Gain} = \frac{-Rf}{Rin}$$ Where Rf = feedback resistor value and Rin = input resistor value

The minus sign in the formula indicates that the output will always be inverted in relation to the input. This means that if the input is positive, the output will be negative and vice versa.

Example

When Rf = 100K and Rin = 10K

$$\text{Gain} = \frac{-Rf}{Rin} = \frac{-100K}{-10} = -10$$

(Note that the gain has no units – simply a mathematical value.)

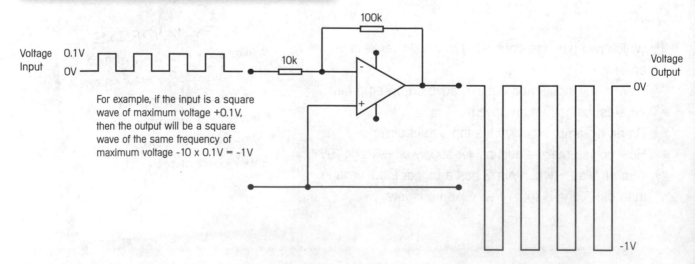

For example, if the input is a square wave of maximum voltage +0.1V, then the output will be a square wave of the same frequency of maximum voltage -10 x 0.1V = -1V

Interfacing a 555 IC and an Op-Amp

Interfacing a 555 IC and an Op-Amp

The circuit diagrams show an op-amp operating as a voltage comparator using a 9V single power supply.

Op–amp and 555 IC Monostable

The op-amp is triggering the 555 IC monostable circuit by taking pin 2 to low voltage, i.e. less than one-third of the battery voltage.

Interfacing means joining two circuits together.

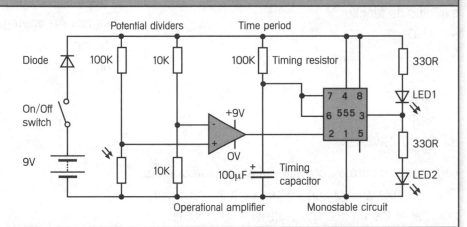

Op–amp and 555 IC Astable

The op-amp is controlling the 555 IC astable circuit by providing a voltage to pin 4 of the astable.

When pin 6 of the op-amp is high, the LEDs flash. When pin 6 of the op-amp is low, pin 4 on the 555 IC resets the output back to 0V and LED1 stays lit.

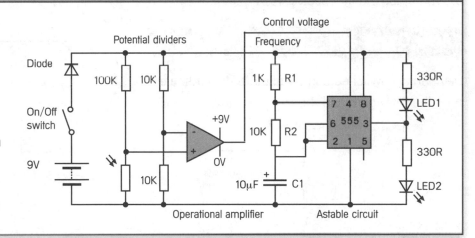

Quick Test

1. What is the function of potential dividers in an op-amp circuit?
2. Write down the formula for calculating the gain of an inverting op-amp.
3. What does Rf stand for?
4. What does Rin stand for?
5. What does the minus sign in the gain formula tell you about the output?

KEY WORDS
Make sure you understand these words before moving on!
- Open loop
- Inverting amplifier
- Gain = -Rf ÷ Rin
- Interfacing

Logic Gates

Logic Gates

Logic gates are digital electronic devices that have several inputs and one output. They're used to make decisions based on the condition of signals at the inputs. The name of each type of logic gate explains its function:

- AND
- OR
- NOT
- NAND
- NOR
- EXCLUSIVE OR

Logic gates operate according to **strict logical rules**. Their outputs only change when certain conditions are met at the inputs. The state of the output is controlled by the state of the inputs and the function or type of logic gate used.

Logic gates are **digital devices** so their inputs and outputs will only ever be at **LOGIC 1** (High) or **LOGIC 0** (Low).

The input signal to a logic gate has to be a **digital signal**. An **analogue signal** that's constantly changing by small amounts can confuse an electronic system that includes logic gates.

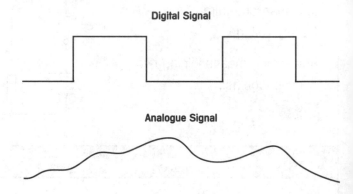

Digital Signal

Analogue Signal

AND Gates and OR Gates

Mechanical switches can be used to demonstrate how the AND gate and the OR gate function.

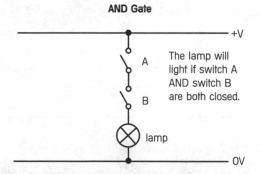

AND Gate

The lamp will light if switch A AND switch B are both closed.

lamp

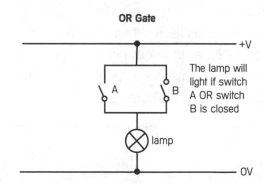

OR Gate

The lamp will light if switch A OR switch B is closed

lamp

DIL ICs

Logic gates can be made from discrete components, but they're usually made as **DIL ICs**. They normally have a number of the same logic gates in the IC package. The gates operate separately, but get their power from the same power supply as the IC.

The diagram shows a **14 pin DIL package**, which contains four NAND gates.

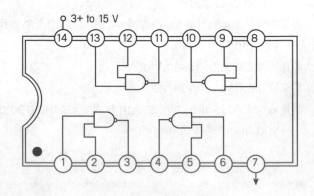

Symbols and Truth Tables

Truth tables list all the possible combinations of inputs and the resulting outputs.

The Two-Input AND Gate

The output Q will be 1 when the inputs A and B are both 1.

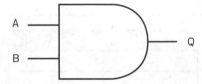

A	B	Q
0	0	0
0	1	0
1	0	0
1	1	1

The Two-Input OR Gate

The output Q will be 1 when the inputs A or B are 1 or both inputs are 1.

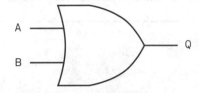

A	B	Q
0	0	0
0	1	1
1	0	1
1	1	1

The NOT Gate

The NOT gate **inverts** (changes around) the signal. The output Q will be 1 when the input A is 0 and vice versa.

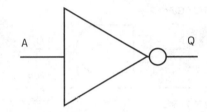

A	Q
0	1
1	0

The Two-Input NAND Gate

The truth table for a NAND gate is the opposite of the AND gate.

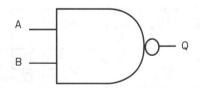

A	B	Q
0	0	1
0	1	1
1	0	1
1	1	0

The Two-Input NOR Gate

The truth table for a NOR gate is the opposite of the OR gate.

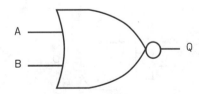

A	B	Q
0	0	1
0	1	0
1	0	0
1	1	0

The Two-Input Exclusive OR Gate (XOR)

The Exclusive OR gate is a true OR function. The output Q will be 1 when the inputs A or B are 1.

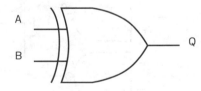

A	B	Q
0	0	0
0	1	1
1	0	1
1	1	0

Universal Building Block

The NAND logic gate is known as the **universal building block** because every type of logic gate can be made by combining NAND gates.

The advantage of using only NAND gates to construct logic circuits is that it reduces the number of different logic ICs in a circuit.

This results in...
- a smaller PCB
- reduced cost
- less soldering
- less need to stock a wide range of logic ICs.

Combinational Logic — Edexcel • OCR

NOT Gate

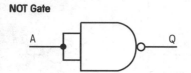

A	Q
0	1
1	0

AND Gate

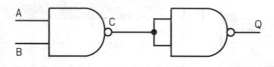

A	B	C	Q
0	0	1	0
0	1	1	0
1	0	1	0
1	1	0	1

OR Gate

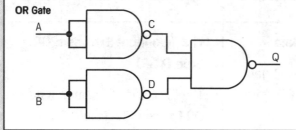

A	B	C	D	Q
0	0	1	1	0
0	1	1	0	1
1	0	0	1	1
1	1	0	0	1

NOR Gate

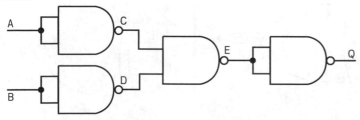

A	B	C	D	E	Q
0	0	1	1	0	1
0	1	1	0	1	0
1	0	0	1	1	0
1	1	0	0	1	0

Exclusive OR (XOR) Gate

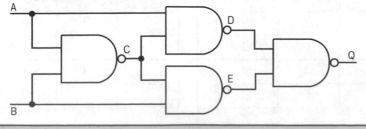

A	B	C	D	E	Q
0	0	1	1	1	0
0	1	1	1	0	1
1	0	1	0	1	1
1	1	0	1	1	0

Circuits Made from NAND Gates

Astable Circuits

Astable circuits with an equal mark–space ratio can be made by using two NAND gates as **NOT gates**, connected in **combination** with a non-polarised capacitor and a fixed or variable resistor.

The value of C, the **non-polarised capacitor**, needs to be small (around 100nF) and the resistor needs to have a **high resistance value** of around 1M5.

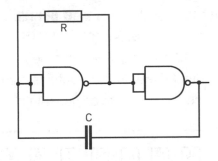

Time Delay Circuits

Time delay circuits can be made by using two NAND gates as NOT gates in combination with a **resistor / capacitor timing network** connected to Gate A. As the capacitor charges, the logic state on Gate A changes from logic 0 to logic 1. With a 9V supply, this would happen at 8V.

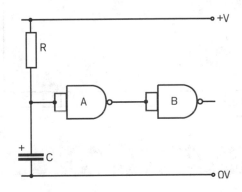

Latch Circuits

Latch circuits can be made by using two NAND gates as NOT gates in combination with the output of Gate B connected to the input of Gate A by a **fixed resistor**. If the logic state changes at Gate A, it will not change the logic state at the output of Gate B due to the **feedback resistor**, which should have a value of around 100K.

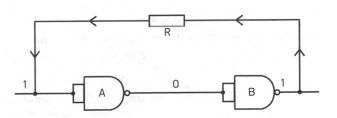

Quick Test

1. Why is the NAND logic gate often referred to as the universal building block?
2. True or false – two NAND gates connected together can make an AND gate.
3. True or false – three NAND gates connected together can make an OR gate.
4. True or false – other types of electronic circuits can be made from NAND gates.

4017 IC Decade Counter

4017 IC Decade Counter AQA • OCR

The **4017 IC** is available in a **16 pin DIL package** and can operate between 3V to 15V. The IC consists of **10 outputs,** which can be reset back to the first output if less than 10 outputs are needed. Only one output is high at any one time.

If LEDs are connected to the output pins, they will each go high in their set order. The speed at which the LEDs turn on and off will depend on the speed of the **pulse generator** or **clock**, which is connected to the **decade counter**.

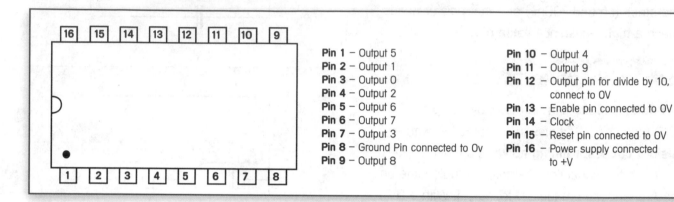

Pin 1 – Output 5
Pin 2 – Output 1
Pin 3 – Output 0
Pin 4 – Output 2
Pin 5 – Output 6
Pin 6 – Output 7
Pin 7 – Output 3
Pin 8 – Ground Pin connected to 0v
Pin 9 – Output 8

Pin 10 – Output 4
Pin 11 – Output 9
Pin 12 – Output pin for divide by 10, connect to 0V
Pin 13 – Enable pin connected to 0V
Pin 14 – Clock
Pin 15 – Reset pin connected to 0V
Pin 16 – Power supply connected to +V

Decade Counter Example AQA • OCR

The circuit diagram shows a **counting circuit** using a 555 IC Monostable with a **short time delay** clocking into a 4017 IC decade counter. The circuit is being used as a **steady hand game** and after five lives have been lost, the thyristor latches and the buzzer sounds.

The monostable, set on a short time delay, is operating as a type of **Schmitt Trigger** to eliminate **switch bounce**.

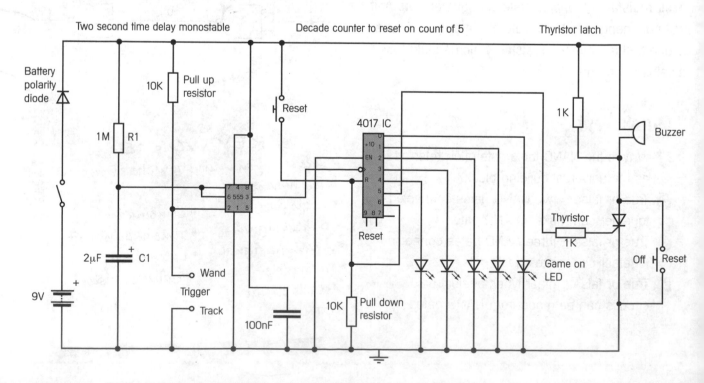

4026 IC Decade Counter

4026 IC Decade Counter
AQA • OCR

The **4026 IC** is a decade counter with a built-in **seven-segment decoded output**.

The 4026 IC is available in a **16 pin DIL package** and can operate between 3V and 15V. The IC counts clock pulses and displays the output as a decimal number between 0 and 9 and on a **seven-segment display**.

Using a 4026 IC to control a seven-segment display avoids using a binary-coded decimal counter IC and a seven-segment decoder IC.

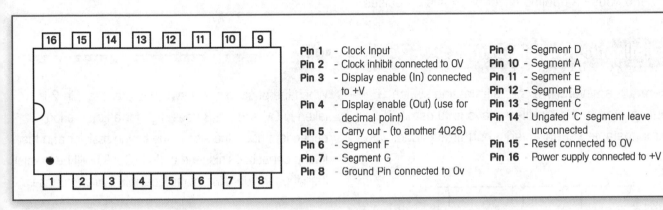

Pin 1 - Clock Input
Pin 2 - Clock inhibit connected to 0V
Pin 3 - Display enable (In) connected to +V
Pin 4 - Display enable (Out) (use for decimal point)
Pin 5 - Carry out - (to another 4026)
Pin 6 - Segment F
Pin 7 - Segment G
Pin 8 - Ground Pin connected to 0v

Pin 9 - Segment D
Pin 10 - Segment A
Pin 11 - Segment E
Pin 12 - Segment B
Pin 13 - Segment C
Pin 14 - Ungated 'C' segment leave unconnected
Pin 15 - Reset connected to 0V
Pin 16 - Power supply connected to +V

PIC Microcontroller
AQA • OCR

The diagram shows a PICAXE 08M microcontroller with one Input and four outputs controlling two 4026 ICs.

The microcontroller is programmed to send a pulse into the 4026 IC (on the right of the diagram) each time the push to make switch is pressed. The circuit only uses two of the microcontroller outputs. The **clock pulses** are provided by output 0 and a **programmable reset** by output 1.

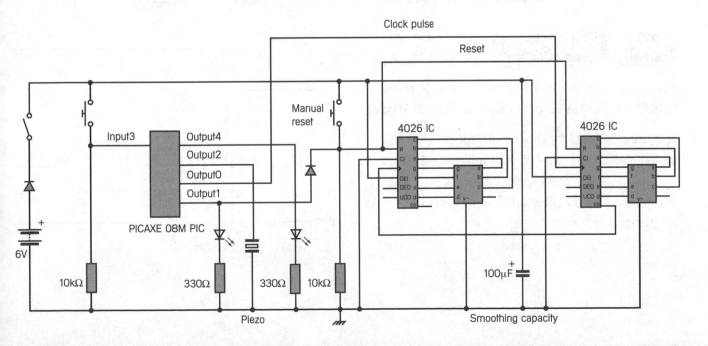

Switch Bounce

Switch Bounce AQA

Switch bounce sometimes occurs when mechanical switches are used in electronic circuits.

A mechanical switch has small **metal contacts** inside it, which come together to make a connection when the switch is closed. When the contacts are nearly touching, current can sometimes jump across the small gap and cause a rippling effect of on and off pulses.

Unwanted pulses can also occur when the contacts come together and bounce back and forth rapidly when the switch is closed. In a counting circuit, switch bounce can cause a single count, made by a switch, to be registered as several counts.

Eliminating Switch Bounce AQA

There are a several ways of eliminating switch bounce. One of the simplest ways is to use a 555 IC monostable set on a very short time period.

When the push to make switch is pressed, pin 2 is taken to OV and pin 3 goes high for a time period, dependent upon the size of the timing resistor and the timing capacitor. This time period (C x R) will be greater than the total duration of the possible switch bounce.

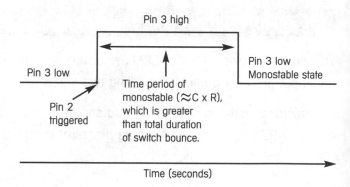

Schmitt Trigger Switch De-bouncing Circuit AQA

Another way of de-bouncing a switch is to use a special type of NAND gate called a **Schmitt Trigger**.

The circuit symbol for a Schmitt Trigger is the same as the NAND gate symbol, with a **hysteresis symbol** (⎍) in the centre of the gate. When the switch is closed, switch bounce is prevented by the discharging action of the capacitor.

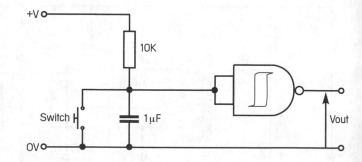

Optoisolators

Optoisolators — AQA

Noise created by motors can damage a PIC microcontroller, which can be overcome by using an **optoisolator**.

Inside an optoisolator is an **infra-red light emitting diode** and a **phototransistor**. The PIC microcontroller is isolated from the motor it is controlling by the barrier between the infra-red LED and the phototransistor. The TLP504A contains two optoisolators in a DIL 8 pin IC.

ICs containing phototransistors are used in switching circuits.

Single Optoisolator

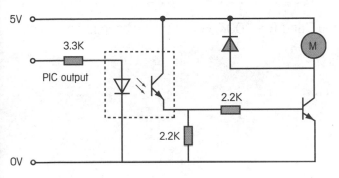

Two Optoisolators In a 8 Pin IC

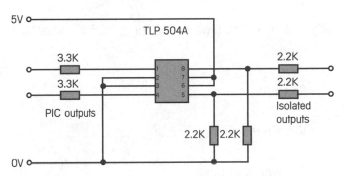

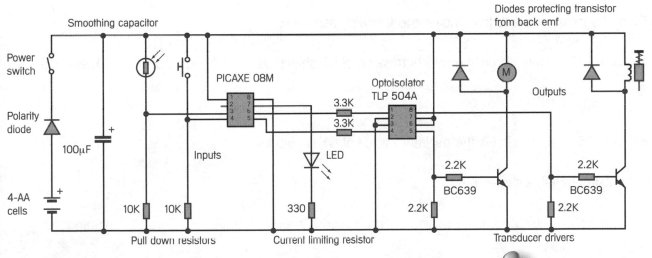

Quick Test

1. What is the working voltage range of a 4017 IC and the 4026 IC?
2. How many outputs does a 4017 IC decade counter have?
3. On a 4017 IC, which pin is the re-set pin?
4. What is switch bounce?
5. How can switch bounce be eliminated?
6. What is the special type of NAND gate that can be used to de-bounce a switch?

KEY WORDS
Make sure you understand these words before moving on!
• Pulse generator
• Schmitt trigger
• Seven-segment display
• Optoisolator
• Phototransistor

Practice Questions

1 Give two uses for a 555 IC working as a monostable.

a) _____

b) _____

2 Give two uses for a 555 IC working as an astable.

a) _____

b) _____

3 Calculate the time period for a 555 IC when C = 100μF and R = 100K. Show your workings and give the units.

Formula: _____

Working: _____

Answer with units: _____

4 Fill in the missing words to complete the following sentences.

a) In a 555 astable circuit, the minimum value of R1 should be _____K and the

maximum value should be _____M.

b) In a 555 astable circuit, the minimum value of C1 should be _____pF and the

maximum value should be _____μF.

c) In a 555 astable circuit, the minimum time period is about _____μS and the

maximum time period is about _____S.

5 Choose the correct word from the options given to complete the sentences below.

digital	**comparator**	**op-amps**	**analogue**

a) Operational amplifiers are often referred to as _____ for short.

b) Operational amplifiers can work as a _____ or an inverting amplifier.

c) Operational amplifiers can be _____ to _____ converters.

6 Calculate the voltage gain of an operational amplifier working as an inverting amplifier when Rf = 100K and Rin = 10K. Show your workings.

Formula: ....

Working: ....

....

Answer ....

7 Draw and label the circuit symbols for the following logic gates.

a) AND Gate	**b)** OR Gate	**c)** NAND Gate	**d)** NOR Gate

8 Draw and label the circuit symbol for an Exclusive OR Gate (XOR) and complete the truth table.

XOR

A	B	Q
0	0	
0	1	
1	0	
1	1	

9 Give a reason why the Exclusive OR Gate was developed, by comparing it against the OR Gate.

....

....

10 Name three different types of electronic circuits that can be made from NAND Gates.

a) ....   **b)** ....   **c)** ....

PIC Microcontrollers

Introduction to PICs

A **PIC microcontroller**, also known as **a computer in a chip**, is a special type of IC that can be programmed to respond to input devices and control output devices.

PIC microcontrollers are widely used and can be found in stereo equipment, DVD players, mobile phones, toys, computer products and alarms.

Modern cars can contain around 40 PICs, which can be used for many purposes, including…
- controlling **engine management**
- controlling the **temperature** in the car for passenger comfort.

A washing machine can have one or two PICs that control the whole washing cycle. A microwave oven can have a single PIC that processes information from the keypad and controls various devices within the oven.

A single PIC can replace many components, so manufacturers require **smaller stock levels** and assembly time can be reduced. If a manufacturer wants to change a product, they can alter the programme without re-designing the PCB and changing the components.

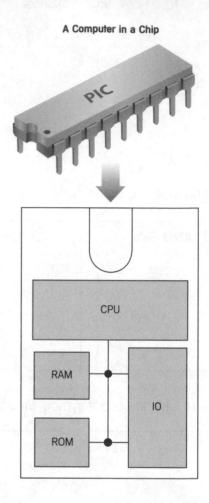

A Computer in a Chip

High and Low Level Programming

In industry, PIC microcontrollers are programmed in a language called **Machine Code** or **Assembler Code** (**low level programming language**).

The **advantage** of low level programming language is that the finished programme is concise, runs faster and is more efficient.

The **disadvantage** of low level programming language is that it's hard to learn and understand.

High level programming languages are used in schools, for example…
- **Logicator** for PIC Micros, and PICAXE (which use flow charts)
- **BASIC** (Beginners All-purpose Symbolic Instruction Code).

Both ways of programming use the same BASIC language commands. BASIC is a **text-based** language used to programme everything from PICs to personal computers.

The **flow chart method** provides a user-friendly graphical way of joining BASIC language commands together and reduces the need for keyboard skills. Flow chart programming uses a smaller number of BASIC commands and is favoured in education.

Complicated and large programmes are best written in BASIC, which has a greater number of commands and is a **more powerful** way of programming PIC microcontrollers.

The disadvantage of high level programming language is that it requires **more** of the PIC's memory and runs slower.

Batteries

PIC microcontrollers run programmes at voltages between **3 and 5.5V d.c**. The d.c. supply voltage can be provided by any of the following:

- 3 x AA **alkaline** cells (1.5 x 3 = 4.5V).
- 4 x **rechargeable** AA cells (1.2 x 4 = 4.8V).
- 5V regulated from 9V d.c. **regulated supply** (5V).
- 4 x AA alkaline cells **in series** with a diode (6V − 0.7 = 5.3V).

Regulated Low Voltage Power Supply

The 9V d.c. supply must be regulated to 5V using a **voltage regulator**. This can be achieved using the L7805 (1A capability), or the 78L05 (100mA capability). The full regulation circuit is shown below. The IN4001 diode provides **reverse connection protection** and the capacitors help **stabilise** the 5V supply.

Input   0V   Output   Output   0V   Input

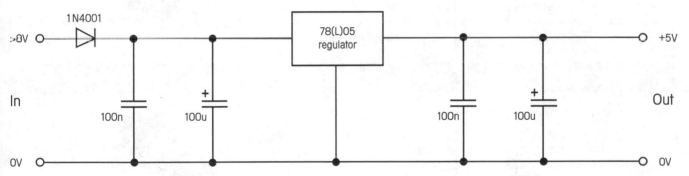

Quick Test

1. Why are PICs known as 'a computer in a chip'?
2. Name three household products that may contain a PIC.
3. What is the difference between a low level and high level programming language?
4. What is the working voltage range of a PIC?
5. Why should a PP3 battery never be used with a PIC?

KEY WORDS

Make sure you understand these words before moving on!
- PIC microcontroller
- Machine Code
- BASIC
- Flow chart
- Voltage regulator

PICAXE Microcontrollers

PICAXE Microcontrollers

PICAXE PIC microcontrollers are manufactured in a **range of sizes** including…

- 8 pin
- 14 pin
- 18 pin
- 20 pin
- 28 pin
- 40 pin.

All are available in a **Dual In Line package**. Larger PICs have a higher number of input and output pins.

All sizes of PICs use the same **BASIC** language and can be programmed by **flow chart** or **written** in BASIC.

PICs can also vary in the amount of **memory** they have, which determines how big a programme can be **downloaded** into the microcontroller. PICAXE microcontrollers have **Analogue to Digital Conversion** (ADC), **infra-red** and many other facilities built into them.

Popular Microcontrollers

These are the three most popular PICAXE microcontrollers:

PICAXE-08M

	Pin	Pin	
+V	1	8	0V
Serial In	2	7	Out 0 / Serial Out / Infraout
ADC 4 / Out 4 / In4	3	6	In 1 / Out 1 / ADC 1
Infrain / In	4	5	In 2 / Out 2 / ADC 2 / pwn 2 / tune

PICAXE-14M

	Pin	Pin	
+V	1	14	0V
Serial In	2	13	Out 0 / Serial Out / Infraout
ADC 4 / Input 4	3	12	Output 1
Infrain / Input 3	4	11	Output 2
Input 2	5	10	Output 3
Input 1	6	9	Output 4
ADC 0 / Input 0	7	8	Output 5

PICAXE-20M

	Pin	Pin	
+V	1	20	0V
Serial In	2	19	Serial Out
ADC 7 / Input 7	3	18	Output 0 / Infraout
Input 6	4	17	Output 1
Input 5	5	16	Output 2
Input 4	6	15	Output 3
ADC 3 / Input 3	7	14	Output 4
ADC 2 / Input 2	8	13	Output 5
ADC 1 / Input 1	9	12	Output 6
Infrain / Input 0	10	11	Output 7

Flow Chart Symbols

Flow charts show the order in which a **series of commands** are carried out, i.e. the **sequence of events** in which something is controlled.

There are different **specific symbols** for each command, linked together by **arrows** to show the correct sequence of events.

Terminator	Decision	Process	Input / Output
Represents start, restart, and stop	Represents a choice that can lead to another pathway	Represents a particular instruction or action	Represents additions to a particular process

Flow Chart and BASIC

The following programmes will make output 2 of a PIC microcontroller go **high** for one second, then **low** for one second, then **repeat the process** until power is disconnected from the PIC.

Basic	Flow Chart
main: high 2	start
wait 1	high 2
low 2	wait 1
wait 1	low 2
go to main	wait 1

Quick Test

1. What pin sizes are PICs manufactured in?
2. True or false – all sizes of PICs can be programmed in BASIC or by flow chart.
3. What does BASIC stand for?
4. True or false – BASIC is quicker to use than flow chart and gives you a shorter programme.
5. What does ADC stand for?
6. True or false – PICs can be remotely controlled by infra-red.

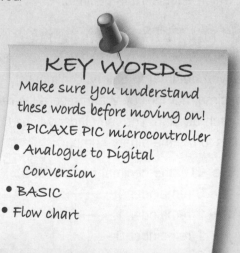

KEY WORDS

Make sure you understand these words before moving on!
- PICAXE PIC microcontroller
- Analogue to Digital Conversion
- BASIC
- Flow chart

Binary Counting

The Decimal System

The everyday **decimal system** of counting uses **ten digits**, 0 to 9. When the count **exceeds** 9, a 1 is placed in a second column to the left of the units column to represent 10s. A third column to the left of the 10s column gives 100s and then 1000s, etc.

The values of the columns starting from the right are: 1, 10, 100, 1000, or in powers of 10: 100, 101, 102, 103, etc.

The Binary System

Digital electronics process **digital signals** that only have **two states**:

- **HIGH**, which equals 1
- **LOW**, which equals 0, therefore it can't use the decimal system of counting, which uses ten digits.

Counting in digital circuits uses the **binary system**, which only has two digits and counts in **powers of 2**. The binary digits 0 and 1 are called **bits** (from the words 'binary digits'). Eight bits make **one byte**. When counting in binary, **many more columns** are needed than when counting in decimal. For example, 11111111 in binary equals 255.

Successive columns from the right represent the binary numbers in powers of 2, i.e. 2^0, 2^1, 2^2, and 2^3, which in decimal would be 1, 2, 4 and 8.

The table shows how the decimal numbers 0 to 15 are coded in binary and require **four bits** to give 1111, which equals 15 in decimal.

The column on the extreme right is the **Least Significant Bit** (LSB) and the column on the extreme left is the **Most Significant Bit** (MSB).

The Binary System of Counting				
MSB			LSB	
2^3 (8)	2^2 (4)	2^1 (2)	2^0 (1)	Decimal
0	0	0	0	0
0	0	0	1	1
0	0	1	0	2
0	0	1	1	3
0	1	0	0	4
0	1	0	1	5
0	1	1	0	6
0	1	1	1	7
1	0	0	0	8
1	0	0	1	9
1	0	1	0	10
1	0	1	1	11
1	1	0	0	12
1	1	0	1	13
1	1	1	0	14
1	1	1	1	15

Converting Binary and Decimal

To convert **decimal to binary**, continuously divide the decimal number by 2 and record the remainder after each division. The remainder will be 0 or 1 and this forms the binary number.

Example – Let the decimal number be 11:

$11 \div 2 = 5$ remainder 1
$5 \div 2 = 2$ remainder 1
$2 \div 2 = 1$ remainder 0
$1 \div 2 = 0$ remainder 1
decimal 11 = (MSB) 1 0 1 1 (LSB) binary

The following example shows how to convert **binary to decimal**:

Let the binary number be 1011:

$1011 = (1 \times 2^3) + (0 \times 2^2) + (1 \times 2^1) + (1 \times 2^0)$
$1011 = 8 + 0 + 2 + 1$
$1011 = 11$ decimal

Using Binary Numbers

It's sometimes easier to use the binary counting system when working with PIC microcontrollers, especially when trying to control **many outputs** at the **same time**.

The following PICAXE programme demonstrates how to make a **seven-segment display** count from 0 to 9. The % sign tells the PIC microcontroller that

binary is used instead of decimal. The PIC output pins are connected to the segments of the display. This means that all 8 outputs can be controlled at the same time.

Remember to always use a **common cathode type** seven-segment display.

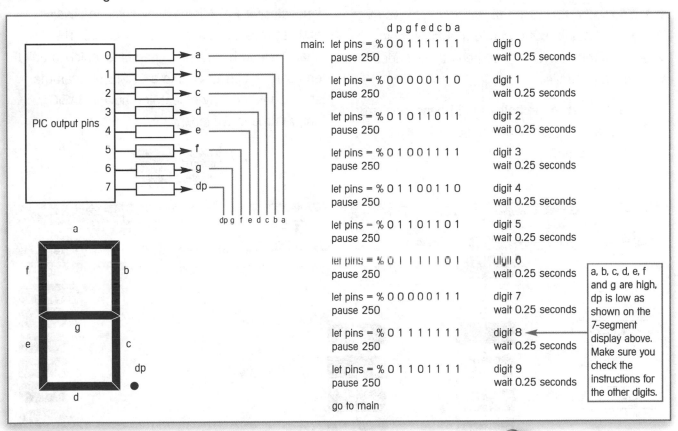

```
                              d p g f e d c b a
              main:  let pins = % 0 0 1 1 1 1 1 1     digit 0
                     pause 250                        wait 0.25 seconds

                     let pins = % 0 0 0 0 0 1 1 0     digit 1
                     pause 250                        wait 0.25 seconds

                     let pins = % 0 1 0 1 1 0 1 1     digit 2
                     pause 250                        wait 0.25 seconds

                     let pins = % 0 1 0 0 1 1 1 1     digit 3
                     pause 250                        wait 0.25 seconds

                     let pins = % 0 1 1 0 0 1 1 0     digit 4
                     pause 250                        wait 0.25 seconds

                     let pins – % 0 1 1 0 1 1 0 1     digit 5
                     pause 250                        wait 0.25 seconds

                     let pins = % 0 1 1 1 1 1 0 1     digit 6
                     pause 250                        wait 0.25 seconds

                     let pins = % 0 0 0 0 0 1 1 1     digit 7
                     pause 250                        wait 0.25 seconds

                     let pins = % 0 1 1 1 1 1 1 1     digit 8
                     pause 250                        wait 0.25 seconds

                     let pins = % 0 1 1 0 1 1 1 1     digit 9
                     pause 250                        wait 0.25 seconds

                     go to main
```

a, b, c, d, e, f and g are high, dp is low as shown on the 7-segment display above. Make sure you check the instructions for the other digits.

Quick Test

1. True or false – decimal and binary are two systems of counting.
2. What are the two digits used in binary counting?
3. In binary counting, what does LSB stand for?
4. In binary counting, what does MSB stand for?
5. How many outputs from a PIC are required to drive a seven-segment display?

KEY WORDS

Make sure you understand these words before moving on!
- Decimal system
- Binary system
- Least Significant Bit
- Most Significant Bit

Programming by Flow Chart

Logicator for PIC Micros

The flow chart programme below makes a **seven-segment display count** from 0 to 9 and repeats the sequence until power to the PIC microcontroller is **switched off**. The PIC microcontroller operates so fast that, without wait commands, the LEDs would switch on and off so quickly that you wouldn't see it happening.

As the flow chart runs, the digital panel shows the **changing states of outputs and inputs** as they would be if the flow chart had been downloaded to a PIC microcontroller.

The **memory bar** in the bottom right-hand corner of the flow chart shows the amount of memory available in the PIC.

The **Comment** command allows you to add short statements to a flow chart. This is useful when using PIC microcontroller programmes in coursework and when answering a PIC question on the written paper.

Comment

Logicator for PIC Micros allows you to develop programmes by flow chart or by BASIC. The flow chart method is easy to understand, quick to build and can be converted into a BASIC programme, showing you a way of learning how a BASIC programme is written.

BASIC

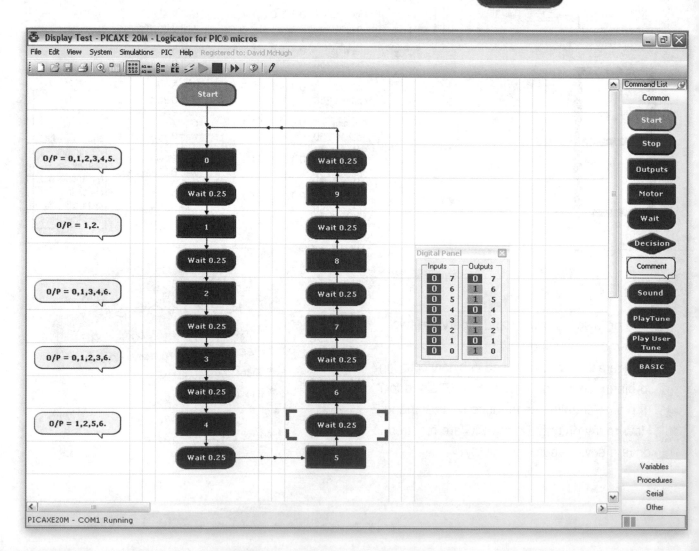

Calibrating Sensors

LDRs and **thermistors** connected to a PIC microcontroller provide information on the conditions they're sensing. This information can be expressed as a number between 0 and 255.

An LDR would give…
- a **high** reading in **bright light**
- a **low** reading in **extreme darkness**.

The **analogue sensor calibrator board** allows analogue sensors to be calibrated so that a switching level can be set within the PIC Programme.

Compare Command

Analogue sensors are used in a control system to inform PIC microcontrollers when to switch output devices on or off when the reading from the sensor reaches a **set switching level**.

When the PIC programme reaches a **compare command**, the programme checks the analogue sensor's reading and compares it with the switching level, which has been set. The programme will continue in the 'Yes' or 'No' direction, depending on the result of the compare.

The programme below will switch lights on if the light level is **less than or equal to 75**.

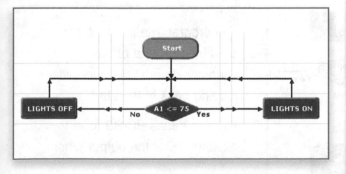

Quick Test

1. True or false – an analogue input device gives a changing signal.
2. True or false – a digital input device gives a signal that is either high or low.
3. What kind of device is a push to make switch?
4. What kind of device is a variable resistor?
5. True or false – a pulsing buzzer is an example of a digital output.
6. True or false – music from a loudspeaker is an analogue signal.

KEY WORDS
Make sure you understand these words before moving on!
- Analogue sensor
- Calibration board
- Compare command
- Less than or equal to 75

Microcontroller Dice Programme

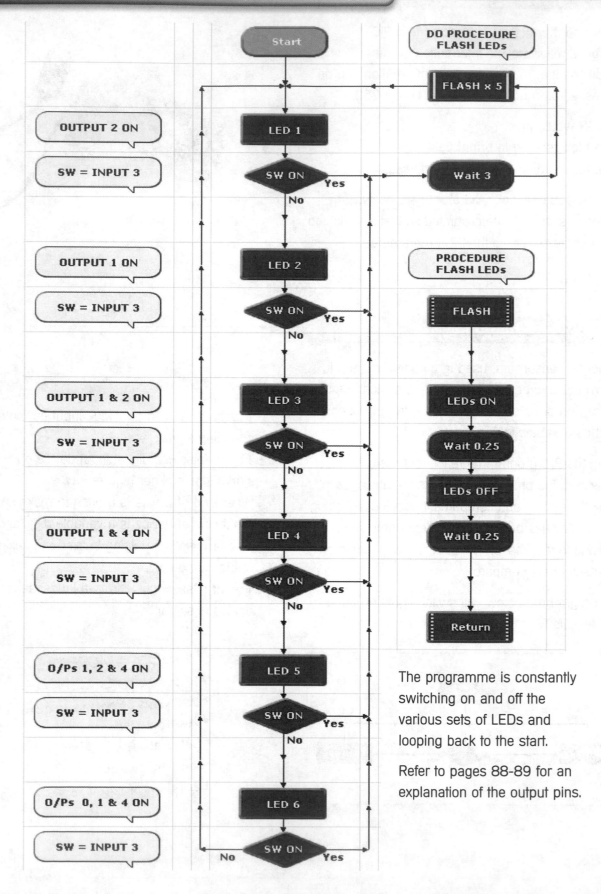

The programme is constantly switching on and off the various sets of LEDs and looping back to the start.

Refer to pages 88-89 for an explanation of the output pins.

Using 1 Input and 4 Outputs

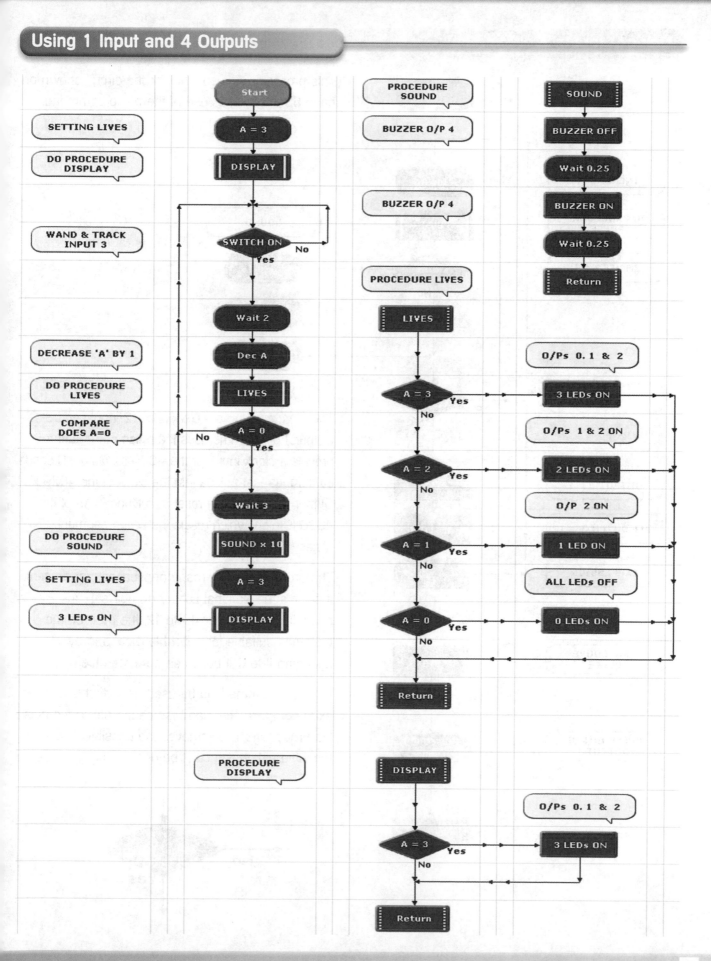

Random Number Programme

This programme is suitable for the circuit shown on page 63 and makes use of the pulse command.

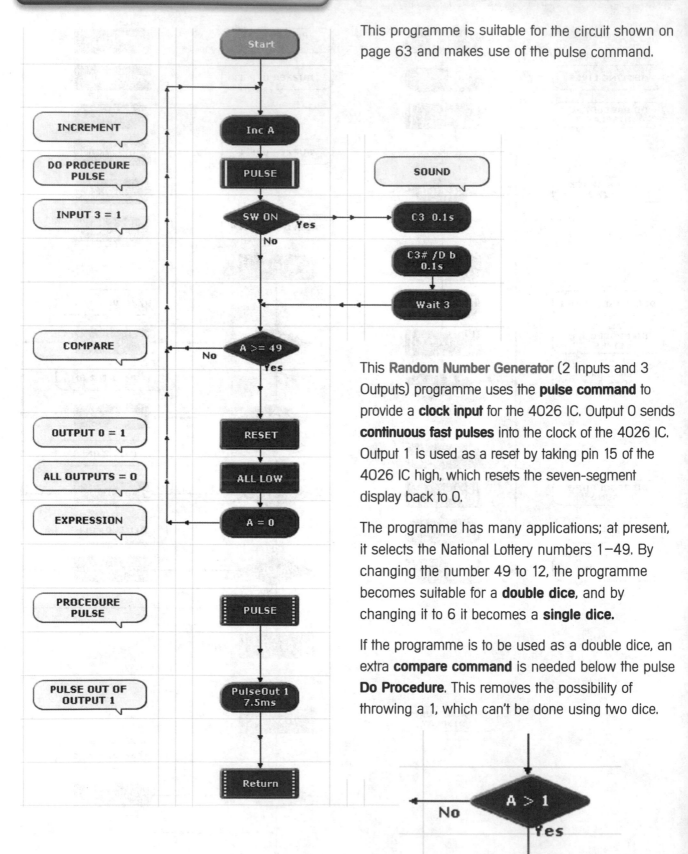

This **Random Number Generator** (2 Inputs and 3 Outputs) programme uses the **pulse command** to provide a **clock input** for the 4026 IC. Output 0 sends **continuous fast pulses** into the clock of the 4026 IC. Output 1 is used as a reset by taking pin 15 of the 4026 IC high, which resets the seven-segment display back to 0.

The programme has many applications; at present, it selects the National Lottery numbers 1–49. By changing the number 49 to 12, the programme becomes suitable for a **double dice**, and by changing it to 6 it becomes a **single dice.**

If the programme is to be used as a double dice, an extra **compare command** is needed below the pulse **Do Procedure**. This removes the possibility of throwing a 1, which can't be done using two dice.

Infra-red Remote Control

AQA

Infra-red

The most popular **remote control** technology in the home is **Infra-Red** (IR), from the Latin meaning 'below' red. The reference to red is the colour of the **longest wave length of visible light**.

The basic concept with infra-red is the use of **invisible light** to carry signals between the remote control and the electronic system that it's controlling.

Transmitting Signals

AQA

An infra-red remote control sends out infra-red **light pulses** that represent specific binary numbers. These binary numbers correspond to commands such as 'volume up', for example, on a TV.

The **infra-red receiver** in the electronic system decodes the pulses of light into binary numbers that the microcontroller can understand. The microcontroller then carries out the command.

The Process of Infra-red Control

AQA

Pushing a remote control button sets off a series of events that cause the **controlled electronic system** to carry out a **command**:

1. The 'volume up' button on the remote control is pressed, causing it to touch the **contacts** beneath and complete the circuit on the circuit board. The integrated circuit detects the **command**.
2. The integrated circuit sends out the **binary number** for the 'volume up' command to the LED.
3. The **LED** sends out a series of **infra-red light pulses** that correspond to the binary 'volume up' command.

Infra-red remote controls...
- have a range of about **10 metres**
- have to be pointing in the general direction of the appliance they're controlling.

Sometimes a remote control with a powerful infra-red LED can **bounce a signal** off walls and ceilings, but infra-red light signals **can't** pass through walls into another room.

Infra-red Transmitter

AQA

A simple transmitter can be made by replacing the red LED in the Rapid Mini-Light project kit (70-0015) with a high power infra-red LED (58-0112).

Make sure that the two button cells and the LED are connected the correct way around. Check if the LED is switching on by looking at the LED through a digital camera.

The transmitter can be put together using double-sided tape.

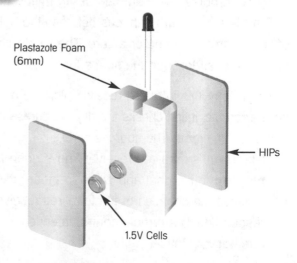

Plastazote Foam (6mm)

HIPs

1.5V Cells

Remote Control Codes

An example of **remote control codes** is the Sony control-S protocol, used in Sony TVs, radios and hi-fi systems. It includes the following **7-bit binary commands**:

Button	Binary Number	Decimal Number
1	000 0000	0
2	000 0001	1
3	000 0010	2
4	000 0011	3
5	000 0100	4
6	000 0101	5
7	000 0110	6
8	000 0111	7
9	000 1000	8
0	000 1001	9
10+	000 1100	12
Channel Up	001 0000	16
Channel Down	001 0001	17
Volume Up	001 0010	18
Volume Down	001 0011	19
Mute	001 0100	20
Power On	001 0101	21

Binary Signals

Sony TV remotes use a **space-coding method** in which the length of the spaces between the pulses of light represent a one or a zero. The length space for a one is **wider** than the space for the zero.

When a Sony TV's **infra-red** receiver picks up a signal from a remote, it first checks the digital **address code** in the **binary signal**. The address code...

- lets the TV's infra-red receiver know when to **respond** to a signal and when to **ignore** it
- reduces the chances of the infra-red receiver responding to a remote control intended for **another appliance**.

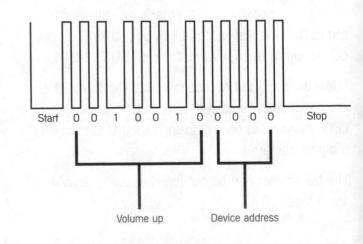

Start 0 0 1 0 0 1 0 0 0 0 0 Stop

Volume up   Device address

Binary Code
AQA

An everyday example of the use of **binary code** is the **barcode**, found on most items in shops.

A barcode is **machine-readable** information, usually of **dark parallel vertical lines** on a white background, representing **ones** and **zeros**. Wrigley's chewing gum was the first product to use barcodes.

Barcodes originally stored data in the widths and spaces of the printed parallel lines, but today they also come as…

- **patterns** of **dots**
- **concentric circles**
- **text codes** hidden in images.

Barcode numbers have different meanings. In the example below…

- the **first six digits**, 137610, are the manufacturer's **identification number**
- the **next five digits**, 74210, are the **product number**
- the last digit – 2 – is the **check code**, which confirms that the barcode was scanned correctly.

Different Types of Remote Control
AQA

The difference between an infra-red remote and the Radio Frequency (RF) remote is that, instead of **light signals**, the RF remote transmits **radio waves** that correspond to binary commands.

RF remote control is used in…
- garage door applications and car alarms
- radio-controlled toys and satellite dishes.

The advantages of RF remotes are that they have a range up to **35 metres** and the radio signals can pass through walls. The disadvantage is the large number of radio signals in the air.

RF remotes overcome **interference problems** by transmitting at **specific frequencies** and using digital address codes in the radio signal.

Manufacturers use **different command codes** in remote controls. Some infra-red remotes can be programmed with **more than one** manufacturer's command code and operate up to 15 different brands of TV, radio, etc. in the same home. SKY satellite television remotes are **universal remote controls** that have to be coded to suit the television receiver they're controlling.

Quick Test

1. Write the binary 00000111 as a decimal number.
2. Write 15 as a binary number.
3. True or false – in a Sony remote control the 'power on' button equals decimal 21.
4. Write 0010100 as a decimal number.
5. Write 16 as a binary number.
6. Write the binary number for the mute button on a Sony remote control.

KEY WORDS
Make sure you understand these words before moving on!
- Remote control
- Infra-red
- Remote control code
- Binary signal
- Binary code
- Barcode
- Universal remote control

PICAXE Microcontroller Infra-red Circuit

AQA

The PICAXE 08M microcontroller has five pins that can be used as **inputs** or **outputs**. The circuit on page 83 has two inputs and three outputs, as shown in the table opposite.

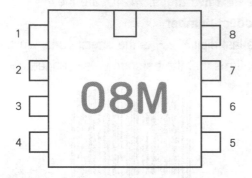

Pin 1	+ Volts
Pin 2	Serial In
Pin 3	Input 4 – Analogue to digital conversion
Pin 4	Input 3 – Infra-red receiver
Pin 5	Output 2 – Piezo sounder
Pin 6	Output 1 – Transistor BC639
Pin 7	Output 0 – LED and also Serial Out
Pin 8	0 Volts

Infra-red Circuit AQA

The infra-red circuit shown below is connected to Input 3, Pin 4, of the PICAXE 08M.

Infra-red Receiver

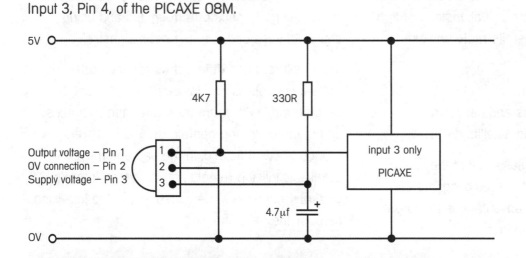

5V

4K7   330R

Output voltage – Pin 1
0V connection – Pin 2
Supply voltage – Pin 3

1
2
3

input 3 only

PICAXE

4.7µf   +

0V

1   2   3

Remote Control AQA

Inside an infra-red receiver is a **photodiode** and a **transistor amplifier** in one package.

When any button on the infra-red remote control is pressed, a binary code is transmitted in the form of infra-red light. This signal takes input 3 low and can be used as a general switch.

If a specific output is to be switched on, extra commands are needed in the program.

Receiver

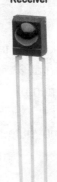

Infra-red Transmitter

PICAXE 08M Infra-red Receiver Circuit

Circuit Diagram

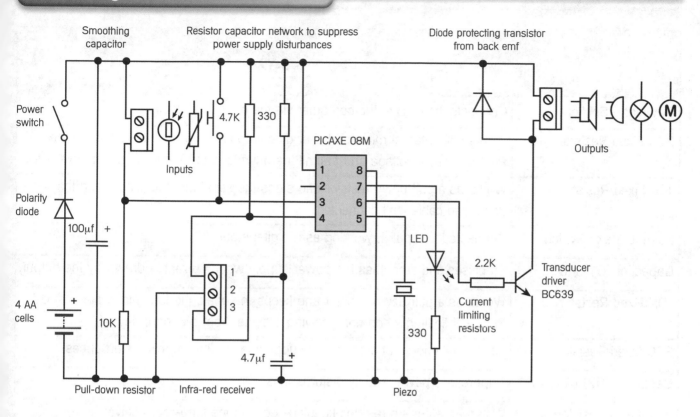

PCB Layout

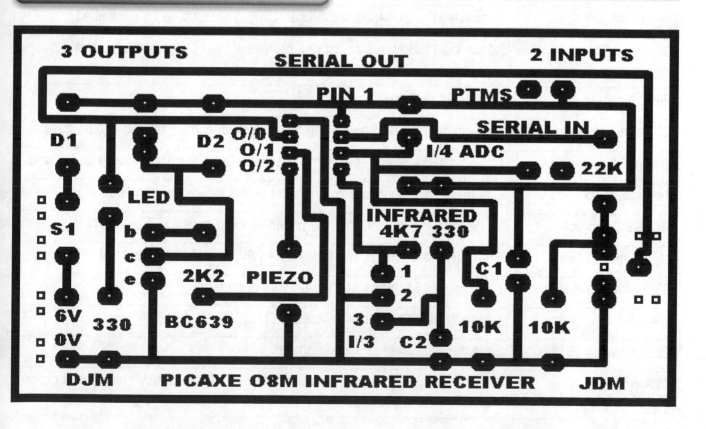

PICAXE 08M Infra-red Receiver Circuit

What the Components Do AQA

Component (looking at the track side of the PCB from right to left)	Component's Role in an Infra-red Circuit
Download Socket	Connects the PCB to the computer via the download cable.
22K Fixed Resistor	Works with internal microcontroller diodes to clamp the serial voltage to the PICAXE supply voltage and to limit the download current to an acceptable level.
10K Fixed Resistor	Works as a pull-down resistor and stops the serial input floating when the download cable isn't connected.
Push to Make Switch	Connected to Input 4, working as a digital input
Capacitor (C1) (100μF)	Stabilises and smoothes the power supply when current is drawn by the output
10K Fixed Resistor	Works as a pull-down resistor and keeps the analogue to digital Input 4 at 0V to stop it floating when not receiving a signal from the input device
330 Fixed Resistor	Limits the amount of current and suppresses power supply disturbances
Capacitor (C2) (4.7μF)	Suppresses power supply disturbances
4K7 Fixed Resistor	Works as a pull-up resistor, keeping Leg 1 of the infra-red receiver high
Infra-red Receiver	Connected to Input 3
Terminal Block	Connected to Input 4, allows different analogue inputs to be connected
8 Pin DIL Socket	Allows the PIC to be removed and protects it from possible heat damage by the soldering iron
Piezo Sounder	Connected to Output 2
2K2 Fixed Resistor	Protects the transistor by limiting the amount of current flowing into its base
Diode (D2)	Protects the transistor against the possibility of back electro-motive force (emf)
Terminal Block	Connected to Output 1, allows different output devices to be used
BC639 Transistor	Works as a transducer driver by providing extra power to the circuit
Light Emitting Diode	Connected to Output 0. Will flicker when PIC is being programmed
330 Fixed Resistor	Protects the LED by limiting the amount of current
Battery Clip	Connects the PCB to four AA cells (6V)
Power Switch (S1)	Switches the power supply on and off
Diode (D1)	Protects the circuit against the possibility of incorrect battery polarity and drops the voltage supply to 5.3V

Infra-red Components List

PICAXE 08M Infra-red Components List — AQA

Quantity	Description	Order Number	Cost per 100
1	Battery Clip (Heavy Duty)	18-0092	6p
1	Battery Holder (4 x AA) Press Stud	18-0115	10p
4	AA Batteries	18-1030	12p
2	10K Resistors	62-0397	1p
1	22K Resistor	62-0405	1p
2	330 Resistor	62-0361	1p
1	2K2 Resistor	62-0381	1p
1	4K7 Resistor	62-0317	1p
1	100µF Capacitor	11-0245	3p
1	4µ7 Capacitor	11-0215	2p
1	BC639 Transistor	81-0080	5p
2	Diode IN4001	47-3130	1p
1	8 Pin IC Socket	22-0150	2p
1	5mm Red LED	55-0155	5p
2	Terminal Block	21-0460	8p
1	Economy Photo Etch PCB	34-0176	50p
1	Miniature Red Rocker Switch	75-0730	28p
1	Miniature PCB Piezo Transducer	35-0215	34p
1	Infra-red Receiver	55-0905	36p
1	Miniature Tactile Switch	78-0630	15p
1	Buzzer	35-0030	49p
1	Uncased Piezo Transducer	35-0200	8p
Optional Components			
1	Miniature Loudspeaker	35-0135	41p
1	Piezo Mini Siren	35-3538	172p
1	Proximity Switch	78-0797	87p
1	LDR	58-0132	65p

Components available from: Rapid Electronics Limited, Severalls Lane, Colchester, Essex CO4 5JS. Telephone: 01206 751166

PICAXE Component List — AQA

Quantity	Description	Order Number	Cost per item
1	PICAXE-08M Microcontroller	AXE-007M	150p
1	PICAXE Stereo Download Socket	CON-039	8p
1	PICAXE Stereo Download Cable	AXE-026	250p

PICAXE components available from: Revolution Ltd, Unit 2 Industrial Quarter, Foxcoat Avenue, Bath Business Park, Bath, BA2 8SF. Telephone: 01761 430044

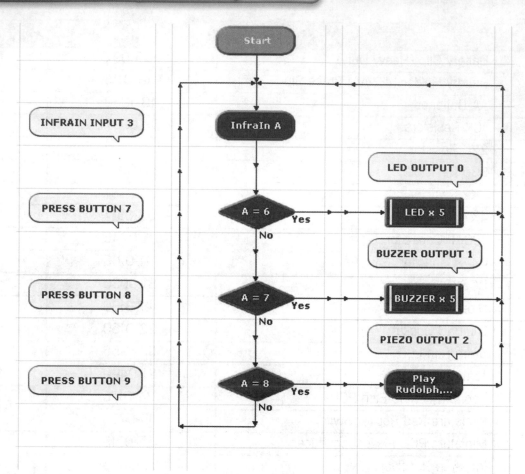

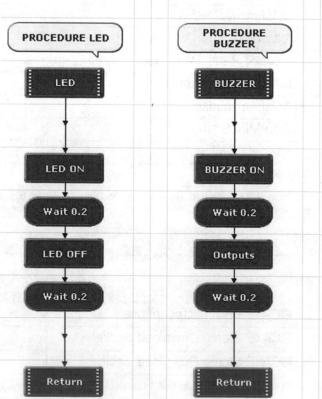

When any of buttons 7, 8 and 9 on the infra-red remote are pressed, outputs 0, 1, and 2 are switched on respectively.

The main programme shows the **InfraIn**, **Compare** and **Do Procedure** commands. The two separate procedures show **Output**, **Wait** and **Return** commands.

Remember that buttons on the remote control are **one number higher** than used in the programme.

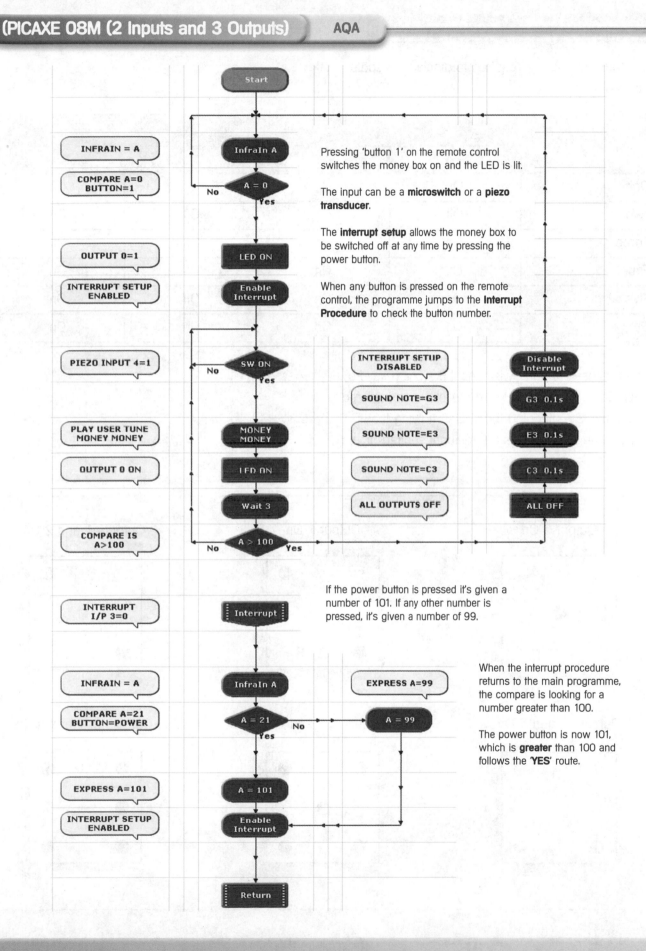

INFRAIN = A

COMPARE A=0
BUTTON=1

OUTPUT 0=1

INTERRUPT SETUP
ENABLED

Start

InfraIn A

A = 0   No

Yes

LED ON

Enable
Interrupt

Pressing 'button 1' on the remote control switches the money box on and the LED is lit.

The input can be a **microswitch** or a **piezo transducer**.

The **interrupt setup** allows the money box to be switched off at any time by pressing the power button.

When any button is pressed on the remote control, the programme jumps to the **Interrupt Procedure** to check the button number.

PIEZO INPUT 4=1

PLAY USER TUNE
MONEY MONEY

OUTPUT 0 ON

COMPARE IS
A>100

SW ON   No

Yes

MONEY
MONEY

LED ON

Wait 3

A > 100   No   Yes

INTERRUPT SETUP
DISABLED

SOUND NOTE=G3

SOUND NOTE=E3

SOUND NOTE=C3

ALL OUTPUTS OFF

Disable
Interrupt

G3 0.1s

E3 0.1s

C3 0.1s

ALL OFF

INTERRUPT
I/P 3=0

Interrupt

If the power button is pressed it's given a number of 101. If any other number is pressed, it's given a number of 99.

INFRAIN = A

COMPARE A=21
BUTTON=POWER

EXPRESS A=101

INTERRUPT SETUP
ENABLED

InfraIn A

A = 21   No

Yes

A = 101

Enable
Interrupt

EXPRESS A=99

A = 99

When the interrupt procedure returns to the main programme, the compare is looking for a number greater than 100.

The power button is now 101, which is **greater** than 100 and follows the 'YES' route.

Return

Electronic Dice (1 Input and 4 Outputs)

Only four outputs are required to display the spots
1 to 6 on a dice.

Number	Outputs			
	2	1	0	4
One	On	Off	Off	Off
Two	Off	On	Off	Off
Three	On	On	Off	Off
Four	Off	On	Off	On
Five	On	On	Off	On
Six	Off	On	On	On

Input 3 is used for the Switch Input (keyboard key F5),
which can be a Push To Make Switch, or an infra-red
transmitter. By attaching the LEDs on flying leads,
they can be arranged as a traditional spotted dice.

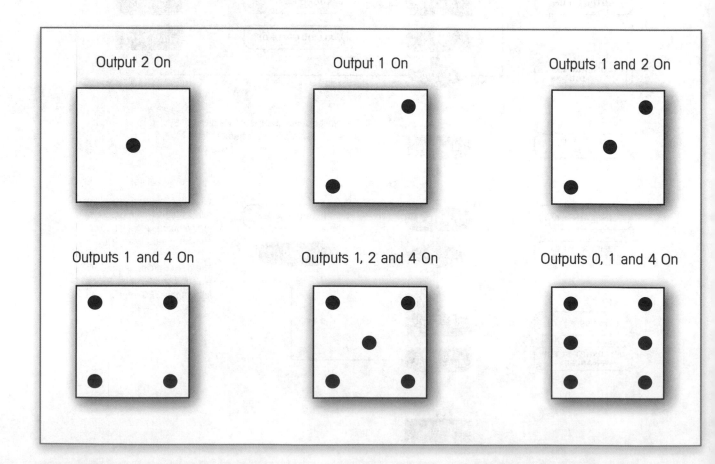

Output 2 On   Output 1 On   Outputs 1 and 2 On

Outputs 1 and 4 On   Outputs 1, 2 and 4 On   Outputs 0, 1 and 4 On

CAXE 08M Microcontroller Infra-red Circuit

Electronic Dice (1 Input and 4 Outputs)

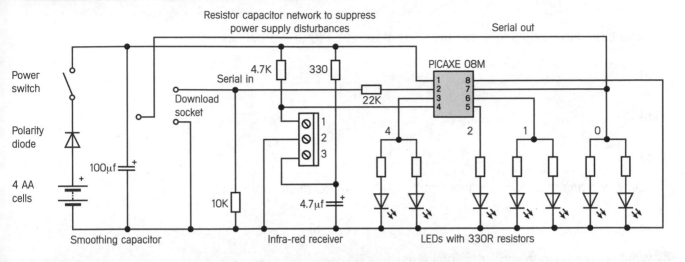

Electronic Dice Programme (1 Input and 4 Outputs)

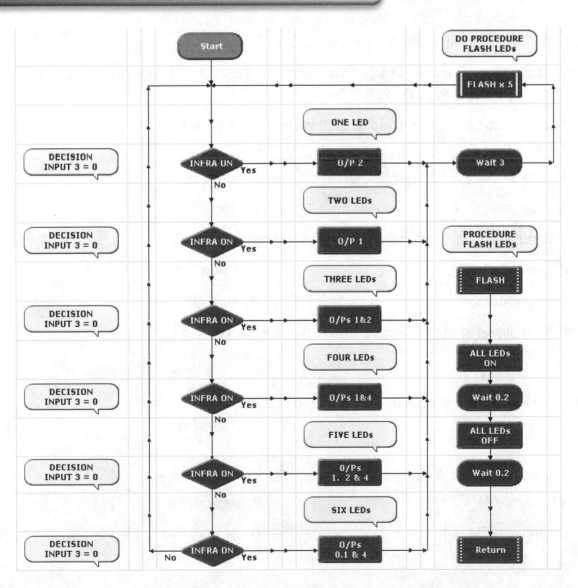

Practice Questions

1. What is a PIC microcontroller?

_____

2. Give three advantages to a manufacturer of using a PIC in a product.

   a) _____

   b) _____

   c) _____

3. Name two analogue sensors that can be used with PIC microcontrollers.

   a) _____   b) _____

4. What is an analogue sensor calibrator used for?

_____

5. Circle the correct options in the following sentences.

   The readings from an analogue calibrator are from 0 to 255. Extreme darkness would equal **0 / 255**. Very bright light would equal **0 / 255**.

6. Which programme command would be used to check the reading from an analogue sensor? Tick the correct option.

   A Procedure ☐   **B** Interrupt ☐

   C Compare ☐   **D** Infrain ☐

7. What does the term infra-red mean? Tick the correct option.

   A Above red ☐   **B** Below red ☐

   C Red message ☐   **D** Through red ☐

8. What do the infra-red signals sent out by a remote control represent?

_____

9. Give an advantage of having a powerful infra-red LED in a remote control.

_____

10. What is the range of an infra-red remote control? Tick the correct option.

    A Around 5M ☐   **B** Around 10M ☐

    C Around 20M ☐   **D** Around 50M ☐

11 Are the following statements **true** or **false**?

a) Infra-red signals will not pass through walls. _____

b) Infra-red light is visible light. _____

c) Sony TV remotes use a space coding method between pulses of infra-red light that represent a one

or a zero. _____

d) A barcode is an example of a binary code. _____

12 Circle the correct option in the following sentence.

The length space for a one is **wider** / **narrower** than the space for a zero.

13 Explain what is meant by the 'address code' in a binary signal.

_____

_____

14 When any button on an infra-red remote control is pressed, how does the PIC microcontroller respond? Tick the correct option.

A The signal takes the input low ◯   **B** The signal takes the input high ◯

15 Draw lines between the boxes to match each of the components in an infra-red circuit with its purpose.

Component	Purpose
330 fixed resistor	Pull-up resistor
4K fixed resistor	Suppresses power disturbances
4.7µF capacitor	Limits the amount of current and suppresses

16 What two components can be found inside an infra-red receiver?

a) _____   **b)** _____

17 Give two advantages of using a radio frequency remote control in place of an infra-red remote control.

a) _____

b) _____

Answers

Material Processes and Practices

QUICK TEST ANSWERS

Page 5
1. Thermosetting plastics; Thermoplastics
2. **Any three suitable examples, e.g:** Acrylic; Polystyrene; ABS; Nylon
3. Acrylic
4. True
5. Medium Density Fibreboard
6. False

Page 9
1. **Any two from:** Polythene; Polystyrene; Polypropylene; Nylon
2. **Any three from:** Split mould; Heater; Hopper; Motor; Hydraulic screw
3. **Any two from:** PVC; Polythene; Polypropylene
4. High Impact Polystyrene

Page 13
1. 4
2. Track side
3. Drilling strain holes to secure flying leads
4. False
5. Red and black

Page 15
1. **Any suitable answer, e.g.** Chair
2. **Any suitable answer, e.g.** Car
3. **Any suitable answer, e.g.** Screws; Nails; Washers
4. False

ANSWERS TO PRACTICE QUESTIONS

Pages 16–17

1.

	Plastic	Timber & Manufactured Board	Metal
Weather resistant	✓		
It work hardens			✓
Has elasticity	✓		✓
Self-lubricating	✓		
Electrical insulator	✓	✓	
Has a grain pattern		✓	
Electrical conductor			✓
Also called deal	✓		

2. **a)–c) Any three from:** Piezo transducer; Smart cable; Optical fibre; Thermocolour sheet; LCD; Smart wire; QTC

3. Heat from the element.
4. **a)–b) In any order:** Through hole method; Surface mount method
5. Tiny bricks with metal end caps.
6. **a)–b) In any order:** Smaller PCBs; Cheaper PCBs
7. By a pick and place component machine.
8. When developing a circuit.
9. **a)–c) Any three suitable answers, e.g.** Reduce the amount of raw materials used; Recycle or re-use materials; Use of biodegradable materials; Toxic waste; Hazardous parts; Amount of waste thrown away; Amount of energy used
10. **a)** Product is designed, made, tested and launched. Sales and profits low. The product is made at a loss.
    b) Sales / manufacturing increase and become profitable. Demand rises; product is selling itself.
    c) Product is at its sales peak and at its most profitable. Competes against new and similar products from rivals.
    d) Sales drop. Manufacturing costs are higher than profits so the product is withdrawn or given a facelift.
11.

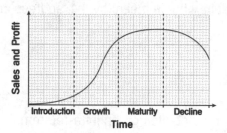

12. **b)–d) Any three suitable answers, e.g.** Using better materials; Adding satellite navigation; Lowering the price; Changing the grille, lights, mirrors, etc.

Basic Principles and Electronic Components

QUICK TEST ANSWERS

Page 19
1. Push to Make Switch
2. Push to Break Switch
3. True
4. Single Pole Single Throw
5. Single Pole Double Throw
6. Double Pole Double Throw

Page 23
1. **Any four from:** AA; AAA; PP3; D; C
2. False
3. 330R
4. Resistor sizes overlap.
5. More resistors in E24; Gold band
6. Single In Line; Dual In Line

Page 25
1. $R_{total} = R_1 + R_2$
2. True
3. False
4. True
5. 2 x 10K, or, 2 x 100K, or any two equal value resistors.

Page 27
1. **a)** About 10M
   b) About 1K
2. Resistance gets smaller as it gets warmer.
3. True
4. Watt
5. 0.25W

Page 29
1. Electrical charge
2. Monostable; Astable; Smoothing

3. The farad is large
4. True
5. Where leads come out of the capacitor
Page 31
1. True
2. Anode and cathode
3. Clamping; Steering; Polarity
4. Light Emitting Diode
5. Short lead; Flat on side of plastic case
Page 33
1. To stop signals feeding back.
2. To protect components.
3. False
4. True
5. To maintain LED brightness.
ANSWERS TO PRACTICE QUESTIONS
Pages 34–35
1. **a)–b) In any order:** Turning the power on and off; Inputting a signal
2. **a)** ○—○ **b)** ○—○
3. B
4. **a) ii)** Blue; **iii)** Grey; **iv)** Orange
   b) Tolerance 5%
5. 350R

6. **a)**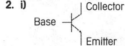
   b)
   c)
   d)

7. **a)**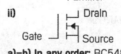
   b)

8. The time it takes the capacitor to charge up to two-thirds of the battery voltage.
9. Five
10. 10 seconds
11. **a)** 10^{-9}
    b) 10^{-12}
12. Allows current to flow
13. **a)** Battery polarity diode
    b) Back emf diode

Transducer Drivers and Electromagnetic Devices

QUICK TEST ANSWERS
Page 37
1. Base; Collector; Emitter
2. True
3. True
4. $hFE = Ic \div Ib$
5. True
6. False
Page 39
1. 220
2. 8800
3. BCX38
4. Field Effect Transistor
Page 41
1. Anode; Cathode; Gate
2. True
3. Electromagnetic devices
4. To interface circuits; To make a latch.
5. **Any three from:** Coil; Core; Contacts; Connections; Armature
Page 43
1. To protect a transistor from back emf.
2. True
3. Stepper; Solar; Toy
4. Stepper motor
5. Reverse the current through the coil.
6. Solder a capacitor to the motor contacts.
Page 45
1. To drive several outputs.
2. **Any three from:** Buzzer; Piezo transducer; Piezo sounder; Loudspeaker
3. False
4. True
5. True

ANSWERS TO PRACTICE QUESTIONS
Pages 46–47
1. **a)** Analogue
   b) Digital
2. **i)** Collector / Base / Emitter
   ii) Drain / Gate / Source
3. **a)–b) In any order:** BC548; BC639
4. **a)** collector
   b) digital
   c) semi-conductor
   d) 0.7
   e) collector; emitter
   f) 1.5V; base
   g) current
   h) 2
   i) collector
5. 400
6. 400.5 mA
7. **a)–b) In any order:** Latch; Transducer Driver
8. Place a 1K fixed resistor in parallel with the buzzer.
9. B; D
10. FETs are voltage controlled and need very little gate current.

Answers

QUICK TEST ANSWERS

Page 49
1. True
2. Astable
3. Monostable
4. With a dot or a notch.
5. Pin 8

Page 51
1. $T = C \times R$
2. Low
3. False
4. Pin 2
5. Timing starts

Page 53
1. Hertz
2. 1 Hertz
3. R1; R2; C1
4. True
5. True

Page 55
1. 4 and 7
2. 2 inputs and 1 output
3. 100 000
4. 6
5. Use two PP3 batteries.
6. True

Page 57
1. To set the reference voltage.
2. Gain = -Rf ÷ Rin
3. Resistor feedback
4. Resistor input
5. Output inverted to input

Page 61
1. All types of Gates can be made from NANDs.
2. True
3. True
4. True

Page 65
1. 3V to 15V
2. 10
3. Pin 15
4. Unwanted pulses.
5. Wait or delay
6. Schmitt Trigger

ANSWERS TO PRACTICE QUESTIONS

Pages 66−67
1. **a)−b) Any two suitable answers, e.g.** Egg timer; Schmitt Trigger
2. **a)−b) Any two suitable answers, e.g.** Motorway flashing sign; Clock
3. 10 seconds
4. **a)** 1K; 1M
   b) 100 pF; 1000 μF
   c) 0.1 μS; 1000 S
5. **a)** Op-Amps
   b) Comparator
   c) Analogue; Digital
6. -10
7. **a)**
   b)
   c)
   d)
8.

A	B	C
0	0	0
0	1	1
1	0	1
1	1	0

9. To give a true OR gate
10. **a)−c) In any order:** Astable; Latch; Time Delay

QUICK TEST ANSWERS

Page 69
1. They contain all the parts of a computer.
2. **Any suitable examples, e.g.** Washing machines; Dishwashers; Microwaves
3. Low level is more powerful and harder to use.
4. 3V to 5.5V
5. PP3 has 9V, which will damage the PIC.

Page 71
1. 8; 14; 18; 20; 28; 40
2. True

3. Beginners All purpose Symbolic Instruction Code
4. True
5. Analogue to digital conversion
6. True

Page 73
1. True
2. Zero and one (0 1)
3. Least Significant Bit
4. Most Significant Bit
5. 8 (including decimal point)

Page 75
1. True
2. True
3. Digital
4. Analogue
5. True
6. True

Page 81
1. 7
2. 00001111
3. True
4. 20
5. 00010000
6. 00010100

ANSWERS TO PRACTICE QUESTIONS
Pages 88–89
1. A programmable IC
2. **a)–c) In any order:** Smaller PCBs; Fewer components needed; Design changes can easily be made
3. **a)–b) In any order:** LDR; Thermistor

4. To express changing conditions as a number
5. 0; 255
6. C
7. B
8. Binary numbers 0 and 1
9. Signals can be bounced off walls and ceilings.
10. B
11. **a)** True
    b) False
    c) True
    d) True
12. wider
13. The address code lets the TV know that the signal is intended for that device.
14. A
15. 330 fixed resistor – Limits the amount of current and suppresses
    4K fixed resistor – Pull up resistor
    4.7μF capacitor – Suppresses power disturbances
16.   **a)–b) In any order:** Photo-diode; Transistor amplifier
17. **a)–b) In any order:** Transmitting range up to 35 metres; Signals can pass through walls

Index

Acknowledgements

The author and publisher are grateful to the copyright holders for permission to use quoted materials and images.
P.5   ©iStockphoto.com/Matthew Silber
P.15   ©2008Jupiterimages Corporation
P.79   ©2008Jupiterimages Corporation

The following images are reproduced with the kind permission of Rapid Electronics Ltd., Severalls Lane, Colchester, Essex CO4 5JS.
www.rapidonline.com (order codes given).
P.10   729000; 631900
P.11   340650; 340500
P.12   855285
P.19   760105; 790295; 781520; 750180; 782845; 780797; 750480; 600520;
        780767; 790100; 780760
P.22   650500; 660105; 670200; 772709; 630010
P.26   580132
P.28   112718; 102216
P.30   473420; 480142
P.31   480142
P.32   480142
P.36   480142; 480142; 810314
P.41   603240
P.48   220130
P.68   473290
P.70   731950; 131224; 731112
P.75   131279
P.79   561250
P.82   130840

With thanks to Revolution Ltd., Unit 2, Industrial Quarter, Foxcoat Avenue, Bath Business Park, Bath, BA2 8SF.

Kitemark reproduced with kind permission of BSI Product Services, www.kitemark.com.

Controlled Assessment Guide
P.6   ©2008Jupiterimages Corporation
P.7   ©2008Jupiterimages Corporation
P.8   Logicator for PIC micros, New Media Learning Ltd and Revolution
        Education Ltd., available from www.picaxe.co.uk
P.8   Circuit Wizard, New Wave Concepts Limited., available from
        www.new-wave.concepts.com
P.10   ©2007Jupiterimages Corporation
        Screenshot from Yenka Technology - www.yenka.com
P.11   Rapid Electronics Ltd
P.13   ©2008Jupiterimages Corporation
        Rapid Electronics Ltd 850695; 850703